Setting Environmental Standards

Setting Environmental Standards

The statistical approach to handling uncertainty and variation

V. Barnett and A. O'Hagan

Department of Mathematics
University of Nottingham
UK

A report to the Royal Commission on Environmental Pollution

CHAPMAN & HALL
London · Weinheim · New York · Tokyo · Melbourne · Madras

Published by Chapman & Hall, 2–6 Boundary Row, London SE1 8HN, UK

Chapman & Hall, 2–6 Boundary Row, London SE1 8HN, UK

Chapman & Hall GmbH, Pappelallee 3, 69469 Weinheim, Germany

Chapman & Hall USA, 115 Fifth Avenue, New York, NY 10003, USA

Chapman & Hall Japan, ITP-Japan, Kyowa Building, 3F, 2-2-1 Hirakawacho, Chiyoda-ku, Tokyo 102, Japan

Chapman & Hall Australia, 102 Dodds Street, South Melbourne, Victoria 3205, Australia

Chapman & Hall India, R.Seshadri, 32 Second Main Road, CIT East, Madras 600 035, India

First edition 1997

© Crown copyright 1997 (year of first publication). This work was commissioned by the Royal Commission on Environmental Pollution and is published with the permission of the Controller of Her Majesty's Stationery Office.

Typeset by AFS Image Setters Ltd, Glasgow

Printed in Great Britain by St Edmundsbury Press, Bury St Edmunds, Suffolk

ISBN 0412 82620 8

A catalogue record for this book is available from the British Library

Library of Congress Catalog Card Number: 97-73539

Contents

Preface   ix
Acknowledgement   xii

1 Introduction   **1**
  1.1 Setting standards: uncertainty and variation   1
  1.2 Terminology   2
    1.2.1 Pollutant   2
    1.2.2 Medium   2
    1.2.3 Location   3
    1.2.4 Sample unit   3
    1.2.5 Sample   3
    1.2.6 Sample statistic   3
    1.2.7 Subject group   4
    1.2.8 Effect   4
    1.2.9 Impact measure   4
    1.2.10 Towards a classification system for environmental pollution   5
  1.3 Overview of environmental standard-setting   5
    1.3.1 Kinds of standard   6
    1.3.2 Balancing cost and benefit   7
    1.3.3 Testing compliance   8
    1.3.4 Where should standards be set?   8
  1.4 Sources of uncertainty and variation   8
    1.4.1 The pollutant–effect relationship   8
    1.4.2 Other causes   9
    1.4.3 Variability within the standard   9
    1.4.4 Sample variation   9
  1.5 Representation of uncertainty   10
    1.5.1 Probability and probability distributions   10
    1.5.2 Interpretations of probability   12
    1.5.3 Risk and risk analysis   13

1.6 Some examples of statistics at work   14
    1.6.1 Diesel fuel   14
    1.6.2 Water distribution and disposal   14
    1.6.3 General   15
1.7 Probability and statistics in standard-setting   15

2 Basic considerations in setting standards   17
2.1 Outline of this chapter   17
2.2 Positioning a standard   18
    2.2.1 Standards as objectives   18
    2.2.2 Positioning the standard   19
    2.2.3 Costs and benefits   20
    2.2.4 Uncertain causality   21
2.3 Ideal and realizable standards   21
    2.3.1 Positioning ideal and realizable standards   21
    2.3.2 Flexibility and complexity   22
2.4 Acknowledging variation   23
    2.4.1 Variation and the statement of an ideal standard 23
    2.4.2 Choice of features to control   25
    2.4.3 Uncertainty due to sampling variation   26
2.5 Statistically verifiable ideal standards   27
    2.5.1 Statistical verification   27
    2.5.2 The statistically verifiable ideal standard   28
    2.5.3 Example   28
    2.5.4 Contrast with ideal and realizable standards 29
    2.5.5 Statistical quality and the benefit of the doubt 30
    2.5.6 Implementation issues   32
2.6 Multiple standards   33
    2.6.1 Uncertainty in the cost–benefit chain   33
    2.6.2 Consistency   34
2.7 Analysing costs and benefits   35
2.8 Current approaches to setting a level   37
    2.8.1 Some current guiding principles   37
    2.8.2 Proper recognition of uncertainty and variation 38
    2.8.3 The precautionary principle   39
2.9 Concluding remarks   40

3 The pollutant–effect relationship and other links   41
3.1 Links in the chain   41
    3.1.1 Links in general   41
    3.1.2 Pollutant and effect   42
3.2 Describing and measuring the pollutant–effect relationship 43
    3.2.1 Mechanistic and statistical models   43
    3.2.2 Hybrid models   44

	3.2.3 Example: radiological protection	45
	3.2.4 Uncertainties in models	46
	3.2.5 Quantifying uncertainty and eliciting expert opinion	47
	3.2.6 Weak links	47
	3.2.7 Model validation	48
	3.2.8 Probabilistic risk assessment and Monte Carlo probabilistic risk assessment	49
	3.2.9 Uncertainty analysis and sensitivity analysis	50
3.3	Recent major reviews	50
3.4	Selective overview of recent published work on pollutant–effect relationships	51
	3.4.1 Sample data	52
	3.4.2 Specific applications	53
	3.4.3 Spatial and temporal variation	54
	3.4.4 Dose-response methods	54
	3.4.5 Bayesian methods	55
	3.4.6 Other statistical methods, and modelling approaches	55
	3.4.7 Deterministic (mechanistic) models	56
3.5	Some current less formal emphases for analysing or categorizing the pollutant–effect relationship	57
	3.5.1 Environmental impact analysis (EIA)	57
	3.5.2 Critical groups	57
	3.5.3 Critical levels and critical loads	58
	3.5.4 Combining information	60
	3.5.5 Extrapolation factors	60
3.6	Summary	61
4	**Current and developing incorporation of uncertainty and variability in standard-setting**	**62**
4.1	Introduction	62
4.2	Statistical pollution studies linked to standards interests	64
4.3	The changing scene: attempts to incorporate statistical arguments in the setting of standards	66
	4.3.1 Contaminated land	66
	4.3.2 Water quality	68
	4.3.3 Air quality	69
	4.3.4 Toxicology and dose response	71
	4.3.5 General matters	73
4.4	Progression	74
5	**Current standards: examples**	**76**
5.1	The extent to which uncertainty and variation are considered	76
5.2	Examples where quite sophisticated treatment of uncertainty and variation is employed	77

5.2.1 Statutory water quality objectives   77
5.2.2 The US ozone standard   79
5.2.3 Land-based disposal of low- and intermediate-level nuclear waste 82
5.3 Examples which pay no regard for uncertainty and variation   86
5.4 Summary   87

6 Conclusions: the current situation and a forward look   **88**
6.1 Broad principles for setting sound standards   88
6.2 The present situation   92
6.3 A look forward   93
6.3.1 Format of standards   94
6.3.2 Setting the level   94
6.3.3 The need for statisticians   95
6.4 Overview   95

References   98
Index   107

Preface

This report has been prepared for and at the request of the Royal Commission on Environmental Pollution. The objective is to provide a critical review of the importance of, and the current state of treatment of, uncertainty and variation in environmental standard-setting. In particular, the report includes:

- an overview of approaches to the treatment of uncertainty and variation in environmental standard-setting, in respect of current practice and future need;
- some examples where uncertainty and variation have been treated fairly well;
- some examples where uncertainty and variation have not been treated well;
- specific proposals for the future approach to setting standards for environmental pollution.

This is the final report; it has been prepared following detailed discussions of an interim report at an international workshop with participants from the European Union and the United States representing different viewpoints and experiences. The final report reflects the constructive and positive outcomes of the workshop.

In presenting this report, there are a few general matters that need to be stressed at the outset and borne in mind as the report is read. First, the Report has been prepared for and at the instigation of the Royal Commission on Environmental Pollution. We trust that it might persuade the Commission of the importance of stressing the need to use statistical (and other appropriate) methods for handling uncertainty and variation in their Report on Setting Standards for Environmental Pollution. Second, standards need to be widely defined: from regulatory conditions endorsed by law, through procedural requirements with economic reinforcements (for example, the 'polluter-pays' principle), to self-imposed rules of behaviour to maintain environmental quality. Aims may range from immediate amelioration of unacceptable conditions, through maintenance of present levels, to sustainability and conservation for the future. Our discussions and recommendations are intended to cover all these forms and aims; the fact that a particular discussion centres on one form and aim or another is purely illustrative and is not intended to be prescriptive or limiting. Third, the report is concerned with general principles and objectives in setting and monitoring standards for environmental pollution in the face of inevitable elements of uncertainty and variation. It is not intended as a comprehensive treatment of the wide range of environmental pollution issues nor does it offer detailed procedures for specific applications. Particular cases are discussed merely to provide examples of the general principles. We believe that urgent need exists for detailed cases to be developed (with appropriate funding) to begin to apply the general principles we have advanced for handling uncertainty and variation in setting standards for environmental pollution.

The report is structured as follows.

Chapter 1   Introduction. This discusses the sources of uncertainty and variation in the setting of environmental standards; establishes terminology; and considers various ways of representing uncertainty that have been proposed, concluding that the appropriate expression of uncertainty and variation is probability and that the proper basis for analysing and interpreting data involving uncertainty and variation is statistical analysis.

Chapter 2   Basic Considerations in Setting Standards. This considers the context of standard-setting in terms of the underlying aims and objectives; contrasts various forms that have been adopted for promulgating standards, with particular reference to the need to recognize uncertainty and variation; relates these to general aims and objectives; proposes a general framework linking (a) the level of pollutant, (b) the formal standard, (c) the effects of pollution, and (d) the criterion for monitoring or testing compliance with the standard; and considers the roles of cost–benefit relationships and formal optimization in the setting of standards.

Chapter 3   The Pollutant–Effect Relationship and Other Links. This considers methods to relate the level of pollution to its effect in terms of impact on some susceptible group, as an example of the many kinds of links along the chain from actions to effects; considers models of such relationships, both statistical and mechanistic; identifies sources of uncertainty in such relationships and highlights appropriate statistical approaches; reviews published work on pollution–effect relationships; and examines, in particular, environmental impact analysis, critical groups, critical loads and levels, combining information, risk analysis, uncertainty analysis, quantifying expert opinion, and extrapolation factors.

Chapter 4   Current and Developing Incorporation of Uncertainty and Variability in Standard-Setting. This extends the interest beyond the pollution–effect relationship to consider statistical investigation that is more specifically related to standards issues; re-emphasizes the forms of standards suitable for incorporation of uncertainty and variational elements; reviews published work involving statistical study of standards; and identifies (from wide international enquiry) some emerging attitudes and principles for incorporation of uncertainty and variation in the setting of standards in different parts of the world.

Chapter 5 Current Standards: Examples. This provides a critical presentation of some examples of current practice in setting standards; and includes examples of standards which employ more thorough (although still incomplete) treatments of uncertainty and variation, examples where uncertainty and variation are recognized but their treatment is flawed and examples where there is apparently no recognition at all of uncertainty and variation.

Chapter 6 Conclusions: The Current Situation and a Forward Look. This draws together the principles and objectives outlined in Chapters 1 and 2, with the developing practices described in Chapters 3 and 4, to produce an overall scheme for incorporating uncertainty and variation in specifying pollution standards and levels for them; recommends a preferred approach to formulating standards, namely statistically verifiable ideal standards; considers the setting of levels for standards and reviews the role of decision theory and other approaches

References. This comprises a list of works mentioned in the text.

V. Barnett
A. O'Hagan
January 1997

Acknowledgement

We are most grateful for helpful comments and information from a large number of individuals and have acknowledged this where appropriate. We hope that we have mentioned all those who have helped, and that we have represented their contributions accurately. However, we accept responsibility and offer our apologies for any omissions in this regard, or any inadvertent misrepresentation.

We were grateful also for the opportunity to discuss our interim report at an international workshop jointly organized by the Royal Commission and the International Centre for Mathematical Sciences and held on 1–3 December 1996 in Edinburgh. Twenty professional workers from the European Union and United States, experienced in the field of environmental pollution and the use of statistical methods, took part in detailed discussions. We appreciate the support given to our report and have sought to represent fully the consensual views that emerged from this workshop.

Finally, we wish to express our thanks to the Royal Commission on Environmental Pollution for giving us this opportunity to contribute to such an important issue as that of setting environmental pollution standards in the face of uncertainty and variation.

V. Barnett
A. O'Hagan
January 1997

Introduction 1

1.1 SETTING STANDARDS: UNCERTAINTY AND VARIATION

Environmental pollution is a major issue of our times. Harrison (1992) is in no doubt that we are facing 'a pollution crisis'. Defining pollution as 'waste that is harmful to other organisms – or to the waste producer', he provides some graphic examples of the scale of the problem: for example, that sea-floor sediment deposits around the UK average 2000 items of plastic debris per square metre.

In seeking to control the problem of pollution, it is necessary to impose requirements on those responsible for the pollution product in an effort to limit its effects on some vulnerable subject group. Ideally, the intention should be to produce some optimal balance between a relevant measure of benefit and an appropriate assessment of cost in this regulatory process. The 'requirements' range in formality from a self-generated 'set of preferred actions' by a commercial organization in support of its quality assurance aspirations, to a specific mandatory (for example, governmental) requirement perhaps expressed in terms of some stated value that a pollutant level should not exceed and the sanctions that will apply in the event of non-compliance. All such 'requirements' can be thought of as *standards for environmental pollution*, and we will throughout employ the term 'standard' broadly to cover all forms of such 'requirement', whether formal or informal, self-generated or imposed, addressed to special groups or the general population, etc.

How are standards set? How should they be set? Clearly, these matters should be discussed in the context of some understanding of the effects of the pollutant. This understanding is often based merely on extrapolation of broad scientific results (for example, 'one part in 100 000 has a critical effect on laboratory animals; better make it one part in 1 000 000 for humans'). For clearer understanding and more objective standard-setting, however, we should declare a specific *aim*, determine what needs to be measured to assess compliance with this aim and then conduct a real-life empirical study of what effect the pollutant has in practice on the specific vulnerable agent. Such knowledge is vital; there can be no point in setting a standard unless we can monitor compliance and assess whether it is achieving its aim.

Thus it becomes essential to understand the real-world effects of pollutants in order to determine how to set standards to control them. Such real-world effects inevitably operate in the face of various forms of uncertainty and variability; they are not fully predictable and have to be expressed in terms of appropriate ways of defining and measuring uncertainty and variation. The effect of a pollutant on a subject will vary for a number of reasons: because of inaccuracies in the measurement or observation process (*observation errors* or *measurement errors*), because effects are not constant over space or time (*spatial* or *temporal effects*); and because of intrinsic differences between one individual and another in their reactions to the pollutant stimulus (*natural variability*).

To understand relationships between a pollutant and an agent in the face of such forms of uncertainty and variation requires probabilistic assessment of levels of uncertainty and statistical methods to infer and explain the relationships. We will present the case that standards cannot be set without such acknowledgement of the central roles of uncertainty and variation and without statistical procedures to handle them.

1.2   TERMINOLOGY

The requirement to set a standard arises from a recognition or suspicion that one or more substances, when present in the environment in sufficiently high levels, will have undesirable, adverse effects of some kind. This is a rather vague statement. In order to discuss the process of standard-setting in sufficient generality, we need to begin by introducing a number of terms which will have technical meanings throughout this report. One benefit of defining terms in this way is to exhibit the variety of standard-setting contexts that fall within the scope of this report.

1.2.1   POLLUTANT

The term *pollutant* will mean the substance or substances whose presence in the environment is of concern. The term might refer to a particular chemical (such as ammonia) or a group of chemicals (such as nitrates or greenhouse gases), an isotope (such as iodine-131) or even an organism (such as *E. coli*). It might refer to a miscellaneous collection of substances (as in standards for bathing water). For convenience we will always refer to 'the pollutant' in the singular, recognizing that in some contexts it can refer to a collection of substances.

1.2.2   MEDIUM

Concern will relate to the presence of the pollutant in some *medium*. The term might be as general as air or water, but will typically be more specific. For example, concern over the presence of ozone as a health hazard might relate to its presence in the atmosphere at ground level, so in this case the medium will be the air up to some defined height above ground. Concern over bathing water quality will define the medium as those waters (typically coastal sea waters) which are to be made safe for bathing and/or water sports.

1.2.3   LOCATION

The term *location* will refer to a specific instance of the medium to which a standard will apply. So the location for a river water-quality standard might be an individual stretch of a river. The standard will generally be set to apply to all locations within the authority of the body setting the standard, but the concept of a location is valuable because we can treat it as the smallest unit to which the standard applies, and at which we can test or verify compliance to the standard.

1.2.4   SAMPLE UNIT

To test or monitor the level of pollutant at a location, it is not normally possible to measure the level in the whole location, so that the usual process is to take samples. A *sample unit* is an individual sample of the medium at a location, taken at some point within the location, and at some point in time.

1.2.5   SAMPLE

A *sample* will be a collection of sample units taken according to some defined scheme or strategy for sampling. For instance, it might comprise a specified number of sample units taken at random points in the location, all at the same time (meaning in practice taken successively over a short time-span). Alternatively, it might comprise sample units taken at a single point at regular time intervals over a specified period.

1.2.6   SAMPLE STATISTIC

The term *sample statistic* will refer to the result of any calculation based on measurements on the sample units in a sample. Again, this is a quite general concept. The measurements made on the sample units will usually be concentrations of the pollutant, but need not be; in some contexts it is uneconomical to measure the pollutant itself, and some surrogate measurements may be made. For example, we might measure levels of nitrogen dioxide in an urban location, or surrogate forms such as incidence of related respiratory illness or even 'predictions' of NO_2 levels based on meteorological and other measures. The measurements might be used, for instance, to calculate an average, or mean, or to count how many sample units have measurements over some threshold, or to calculate the range of measurement values in the sample. Remembering that 'the pollutant' might refer to a collection of substances, it is clear that the measurements might be of various different kinds, and the sample statistic may also be a collection of figures (for instance, the average concentrations of each different chemical in the sample units). So again, the singular term 'sample statistic' is convenient, but in a given context may refer to a *collection of numbers*: to what is termed *multivariate data*. In some circumstances different measures of pollution or effect may be available. There will be an attraction in choosing the least variable measure, provided it appropriately represents what we are seeking to express.

1.2.7   SUBJECT GROUP

We now turn to definitions relating to the effect of pollution. The term *subject group* refers to those individuals or organisms on whom the effect of the pollutant is the basis of concern and the motivation for setting standards. This is another concept which will vary considerably according to context. The subject group might be a collection of people (the general population, or some specific group such as children or radiation workers), of some other animals (such as fish or grazing animals), of plants (such as forest trees or a food crop), etc. In setting standards for river water we might regard fish in the river as the subject group, whose health and population numbers are affected by pollution, or we might focus more narrowly on fishermen as the subject group, who are affected if there are too few of specific species of fish. In some contexts we might have in mind several different subject groups. A group of special interest, perhaps because of its high susceptibility to the pollutant or its high degree of contact with the medium, may be called a 'critical group' (section 3.5.2).

1.2.8   EFFECT

By *effect* we shall mean some measure of the condition (often well-being) of a subject group that is (expected to be) affected by the pollutant. In a broad sense the intention of setting standards is to increase, or prevent or limit any decrease in, the well-being of the subject group. 'Effect' may be a relative measure, of change in well-being from some reference level (in keeping with the common usage of the term), but equally could be an absolute measure of well-being.

1.2.9   IMPACT MEASURE

Just as we distinguish between the level of pollutant at a location and the value of a sample statistic which results from trying to measure or monitor the level of the pollutant, it is sometimes useful to distinguish between effect and the result of trying to measure it. Thus, an *impact measure* will be like a sample statistic, typically the result of measuring effect on only a sample of individuals in the subject group (for example, incidence of illness produced by the environmental pollutant). It may also be a surrogate when the measurement of effect itself is impractical.

1.2.10   TOWARDS A CLASSIFICATION SYSTEM FOR ENVIRONMENTAL POLLUTION

We have defined above the concepts of 'medium', 'location' and 'subject group' to delineate the 'effects' of a pollutant. Essentially, this amounts to the beginnings of a *classification scheme* or *taxonomy* for environmental pollution. It will prove a valuable (and adequate) structure for our subsequent discussion. However, there is much general advantage to be gained from use of a comprehensive taxonomy for pollution. This was remarked by Barnett (1997) in his comprehensive review, for the 1996 SPRUCE conference on *Statistical Aspects of Pollution: Assessment and Control*, of statistical studies of pollution effects published in the last 5 years. He identified in excess of 1000 publications across all forms of pollution, all types of effect, all manner of forms of uncertainty and variability and utilizing all aspects of the wide range of existing (and newly developed) statistical models and methods. The work occurred across all disciplines, including agriculture, biology, engineering, environmental science, hydrology, medicine, meteorology, and statistics *per se*. Only about 10% of the publications were actually in statistical journals.

To understand the complexity of the relationships, Barnett (1977) found it necessary to develop an appropriate taxonomy, which over 12 levels (some crossed and some nested) expressed all relevant definitive aspects of the pollutant, its effects and the studies of these effects.

This classification system for the pollutant *per se* included the medium carrying the pollutant (air, fresh water, sea water, soil, food, solid, etc.), the nature of the pollutant (organic or inorganic, bacteriological or viral, particulate, gaseous, radiation, etc.), the specific forms (salmonella, lead, dioxin, ozone, gamma rays, etc.). Also included were what the pollution was affecting (fauna, flora, humans, ecosystems, etc.), in what specific forms (birds, forests, elderly people, rural communities, etc.), what was the general field of the effect (health, morbidity, strength/stability, sustainability, balance/condition, etc.) and its specific form (cancer, death rates, biofouling, despoilation, output quality, etc.).

1.3   OVERVIEW OF ENVIRONMENTAL STANDARD-SETTING

As is clear from the previous section, the wish to set standards for environmental pollution arises in a great variety of contexts. Although we have introduced some terminology to allow us to embrace this variety succinctly in our discussions, the underlying diversity should be borne in mind throughout, and we shall try to reinforce this by illustrating each point in various ways.

The following overview of practices of standard-setting is intended to 'set the scene' and provide some basis for the discussion of sources of uncertainty and variation which follows. It will be greatly enlarged upon in Chapter 2.

1.3.1   KINDS OF STANDARD

In keeping with the diversity of applications, standards are actually promulgated in practice in a variety of ways (although we must remark with regret the fact that too often standards are advanced with little or no rationale given for their form).

At one extreme, a standard might be set *to place a limit on the level of effect on a subject group*, without actual reference to any pollutant which causes or influences that effect. An example (albeit a rather extreme one) is the UK government's standard for disposal of low- and intermediate-level nuclear waste (see Department of the Environment (DOE) *et al.* 1984). The implication is that appropriate but unspecified measures need to be introduced to limit deaths from radiation escape to a maximum of 10^{-6} per annum.

More commonly, a standard is set on *the level of pollutant in the medium at each of a specified class of locations*. Again, this is often a target which is set without reference to the measures that will need to be introduced in order for it to be achieved, as with the setting of bathing water standards which limit the level of faecal coliforms without saying how the level is to be attained (for instance, by controlling sewage effluents). It is clearly intended that controlling or limiting the pollutant will limit the effect on the subject group, but the standards are often set without reference to explicit targets for the effect, and with only the most cursory analysis of the relationship between pollutant and effect; or of how to measure it!

When the standard is set on locations which represent the places where the pollutant enters the medium, the control measures are often more explicit, such as sewage discharge consents which clearly place a control requirement on operators of sewage treatment works. In such cases, however, although the control measures may be clearer, the link to effect may be more complex. Thus, standards for sewage entering rivers may be set with the intention of limiting pollutants in the river generally, with a view thereby of improving the quality of the river water or of the estuarial waters into which the river discharges, or of nearby bathing beaches. The point is that the standard may be set at locations where pollution arises, but the interest is in reducing pollutant levels at other, more general locations, representing thereby an extra step between the standard and the effect. An interesting example concerns the emission of substances such as sulphur products into the atmosphere, where the standard is set on factory chimney emissions, but where the interest may be in pollution of lakes by acid rain, which corresponds not only to locations very distant from those on which the standards apply but also to locations in a different medium!

A standard may be set on a specific sample statistic. This has the apparent advantage that compliance to the standard can be determined unambiguously. In contrast, compliance to a standard which refers to the level of pollutant in a location cannot usually be determined precisely because we cannot measure the level in the whole location. Compliance testing inevitably requires use of a sample statistic, with the attendant possibility of error — either the standard may be met over the whole location but a poor sample leads to it failing the test, or the standard may not be met over the whole location but a lucky sample leads to it passing the test. This possibility of misrepresentation is removed by a standard which is set directly on a sample statistic. An example is the European Union bathing water standard (EU, 1994), which requires that no more than a specified number of sample units in a sample of given size may have measured levels of pollutant above a stated threshold. There is, however, an attendant disadvantage in such a standard in that the chain from standard to effect is longer. The standard may be set with the intention of limiting the level of pollutant at each location, and this target may be explicit — as in EU (1994) — but again this target may be set without reference to implied targets on the effect, and without careful analysis of the relationship between pollutant and effect. In particular, compliance with such a standard does not in itself imply any specific level for the pollutant 'at large' even in the defined reference region, nor are statistical methods typically specified or applied to infer this level.

1.3.2   BALANCING COST AND BENEFIT

We have stressed, in the above discussion of kinds of standard, the links between the standard and, on the one hand, the effect on the subject group and, on the other hand, the specific actions needed to achieve the standard. Both of these are important because they represent the two sides of the *cost–benefit balance*. The cost of setting standards lies in the actions needed to achieve them, and the benefit lies in achieving improvements in effect. While these two sides are difficult to quantify and to weigh against each other in practice (as will be discussed in Chapter 2), there must at least be some implicit determination of the balance between them in order to set a standard. No matter what kind of standard is used, setting it at a more stringent level will in general improve the effect on the subject group, but will cost more to achieve. Choosing a particular level for a standard implies placing greater or less value on the benefits relative to the costs. Furthermore, attempts to comply with some standard impose further costs *per se* which should also figure in the cost–benefit balance.

It is therefore hard to justify any approach to setting standards which does not attempt, as far as is realistically possible, to identify the likely benefits and the likely costs. Accordingly, we believe that it is vital to recognize (and to analyse) the relationship between pollutant level and effect, and between pollutant levels and the actions which might reduce them. The Royal Commission is undertaking a separate study of cost–benefit considerations for environmental pollution standards.

1.3.3   TESTING COMPLIANCE

Where the standard is not set explicitly on a sample statistic, there is a need to test compliance. This may at first glance not appear to be very relevant to the process of setting the actual standard. It does not, after all, relate *directly* to either the benefits or costs (assuming that sampling costs are trivial compared with the costs to the polluters of achieving compliance). However, it does have a vital *indirect* connection with benefit. The point is that we cannot be sure in such a case (where the standard is not set explicitly on a sample statistic) that the standard will be met. Therefore the benefit is more unsure in such a case, and this added uncertainty will generally be an important factor in balancing cost and benefit.

1.3.4   WHERE SHOULD STANDARDS BE SET?

We have shown that standards can be, and are, set in a variety of positions: on levels of effect, levels of pollutant, on impact, in terms of sample statistics, etc. We explore this diversity in more detail later (for example, in Chapter 2), but it is worth observing at this stage that the choice of position often appears arbitrary. We are not aware of any detailed data-based case studies designed to assess relative merits of different choices in this regard.

1.4   SOURCES OF UNCERTAINTY AND VARIATION

Uncertainty and variation (variability) pervade all aspects of the standard-setting process. The main sources are as follows.

1.4.1   THE POLLUTANT–EFFECT RELATIONSHIP

The consequences, in terms of the effect on the subject group, of any particular level of pollutant in the medium (and at each location) are inherently uncertain. Uncertainty arises in several guises when we try to examine the pollutant–effect relationship. First, there is uncertainty due to the fact that invariably we have imperfect scientific understanding of the mechanisms by which the pollutant influences the effect. In addition, there is uncertainty due to random variation, and this will typically enter in several ways. There is *natural variation* between individuals in the subject group in terms of their reaction to a given exposure to the pollutant. There is also *variation in the levels of exposure* that individuals in the subject group will receive, because there is inherent randomness in the way in which the pollutant is spread throughout the medium and randomness in how individuals come into contact with the medium. Furthermore, to quantify the effect we will have to measure it on individuals, and there will be further uncertainty and variation introduced by any imprecisions in the measurement process (*measurement errors*).

Consider, for example, radioactive contamination from a nuclear power station. Given a specific level of release of radiation (in the form of a particular quantity of a given radioactive isotope), there is:

- random variation in how that release is transmitted through the air to reach individuals and/or scientific uncertainty about the physics of dispersal of such a release;
- random variation in the habits of individuals, with reference to how they might be exposed to the contamination (for instance, variation in the amount of time they spend out of doors in the vicinity of the station);
- scientific uncertainty about how such radioactive substances are processed through the human body and/or random variation between individuals in the rate at which they expel such substances, leading to uncertainty over the effective dose to a person from a given level of contamination;
- both scientific uncertainty and random variation in the chance that an individual receiving a given effective dose will contract a cancer (or some other effect);
- uncertainty in the form of measurement errors when we seek to measure effects on actual individuals.

This example is not at all untypical in the length and complexity of the connections from pollutant to effect, nor in the fact that uncertainty or random variation affect every step in the chain.

1.4.2   OTHER CAUSES

Additional to the uncertainty over the way in which the pollutant itself influences the effect, the effect will typically have other causes, over which we are not exerting any control. Even if pollution standards are effective in reducing changes in the effect due to the pollutant, adverse changes may arise from many other causes. If, for example, we successfully apply a standard on airborne particulates to reduce respiratory illness, the level of respiratory illness may actually rise because of increased effects of other causes which are affected by the control on particulates.

1.4.3   VARIABILITY WITHIN THE STANDARD

Standards will generally apply limits on some quantity such as a pollutant level. Even if the standard is met, there is likely to be considerable variation from time to time, and from location to location, in how far within the limits the true level lies. Applying a standard (successfully) therefore leaves a residual uncertainty about the margin to which that standard is met in practice.

1.4.4   SAMPLE VARIATION

Concentrations of the pollutant will vary in time and throughout the medium at any location. This leads inevitably to sample variation. As discussed in section

1.3.1, even if we know the results of the sampling, in terms of a relevant sample statistic, we cannot know the true level of pollutant in any given location. This compounds the 'variability within the standard' of section 1.4.3, in the sense that when the standard is set on the pollutant level and we test compliance with that standard by reference to a sample statistic, then even if the compliance test is passed the true pollutant level could be beyond the limit imposed by the standard. There is therefore variability not just *within* the standard but also *beyond* it. Measurement error further compounds these uncertainties.

Sample variation also means that there will always be uncertainty about how successful a standard has actually been, if we try to assess the resulting effect by means of a sample-based impact measure.

1.5   REPRESENTATION OF UNCERTAINTY

The lay interpretation of uncertainty involves notions of chance, randomness, risk, hazard and unpredictability. Its formal expression seeks to incorporate these thrusts. Efforts at representation which allow uncertainty to be quantified and analysed have a long history. Throughout, the concept of *probability* has remained paramount as a basis for expressing uncertainty and as the basis on which *statistical methods* for analysing sample data have been developed. Indeed, the application of statistical method nowadays pervades all areas of research and enquiry. It is central to biomedical studies: in agriculture, medicine, pharmaceutics, etc. It plays a crucial role in industry: in finance, production, quality, etc. Social enquiry and market research lean on sample survey methods, regression analyses, etc. Fields as diverse as archaeology, education, philology and public admini-stration all depend on statistical models and methods to aid understanding and policy-making. In all cases the thrust is to extend understanding in the face of uncertainty and variation and to derive cost-effective and defensible actions in such circumstances. We shall illustrate the important role of statistics throughout this report; section 1.6 presents some illustrative examples of applications of statistics in different fields. Environmental study is no different in this respect, and environmental statistics is as important for proper understanding as is the sound application of the principles and methods of chemistry, engineering or physics.

Let us consider some basic aspects of the use of the probability concept.

1.5.1   PROBABILITY AND PROBABILITY DISTRIBUTIONS

It will be helpful here for some readers if we review briefly the representation of uncertainty by means of probability. The probability of an uncertain event E, based on information H, is denoted by $P(E \mid H)$, and is a number between zero and one. Probability zero represents an event that is impossible (or so improbable as to be effectively impossible, such as the probability of guessing the distance from the centre of the dome of St Paul's Cathedral to the centre of the hinge of the knocker on the door of 10 Downing Street, measured in an absolute straight line, to the nearest millimetre). Probability one represents an event that is certain (or so highly probable as to be effectively certain). Values in between represent all the gradations from impossibility to certainty. The probability $P(E \mid H)$ is generally abbreviated to $P(E)$, 'the probability of the event E', when the underlying information base H is understood.

For example, as remarked above, DOE *et al.* (1984) set out a standard for authorizing a facility for disposal of low- and intermediate-level nuclear waste — that the risk to an individual in the long term (possibly thousands of years), after closure and sealing of the disposal site, of death due to escape of radiation from the site, should not exceed 10^{-6} per annum. While 'risk' is not defined in that document, the most natural interpretation of the figure of 10^{-6} is as a probability, the probability of the event (E) that an individual dies in any given year (averaged over many years) due to exposure to radiation from the site.

While in such a context we are concerned with the probability of a *single event*, representation of uncertainty is more often required concerning the value of an *uncertain quantity* (or *variable*). Consider the problem of expressing uncertainty about the value X of the concentration of pollutant in a single sample item, taken randomly from the total medium at some location. It might, for instance, be the concentration of ozone in a sample of air taken at some location in Manchester. We could use probability by asking for the probability that X is less than or equal to a specified value x, and to ask for this probability for all values x in the feasible range. This, plotted against x, is called the *distribution function* of X, and an example is shown in Figure 1.1. The curve necessarily increases from left to right, because, for instance, the probability that X is less than or equal to 2 must be at least as great as the probability that it is less than or equal to 1.

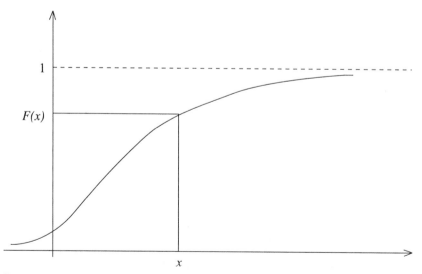

Figure 1.1 The distribution function, $F(x)$.

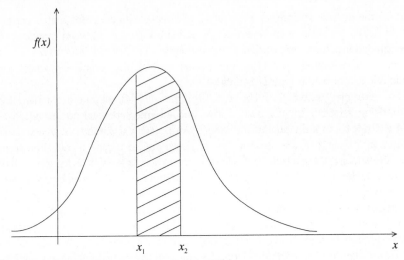

Figure 1.2 The probability density function, $f(x)$.

The distribution function of X is a comprehensive expression of uncertainty about X, but an equivalent representation is more commonly used, which is the *density function* of X. Figure 1.2 shows the density function corresponding to the distribution function in Figure 1.1. Its interpretation is that, given any two values x_1 and x_2, the probability that the true value of X will lie between x_1 and x_2 is the area under the curve enclosed between x_1 and x_2, shown shaded in Figure 1.2. Since X is certain to take *some* value, the area under the whole curve must be one. (Mathematically, the density function is the result of differentiating the distribution function.)

The reason why the density function is preferred is that it gives a convenient visual representation of the *relative* probability of X taking different values. If we place x_1 and x_2 very close together the area under the curve is directly related to its height at x_1 (or at x_2, since they are close together). So the height of the curve at the value x represents the relative probability that X will take a value 'close to' x.

We refer to the formal representation of uncertainty about X, whether by a distribution function or density function, as the *probability distribution* of X (or just 'the distribution of X').

1.5.2 INTERPRETATIONS OF PROBABILITY

Statisticians recognize two different interpretations of the concept of probability. The one that is usually taught, and will be familiar to many readers, is the frequency interpretation. According to the frequency interpretation, a probability $P(E) = \frac{1}{4}$, say, means that out of a long series of occasions on each of which E may or may not occur, it will in fact occur on one-quarter of them and not occur on three-quarters. This interpretation is quite suitable for the examples discussed in section 1.5.1: the probability of 10^{-6}, for death in any year of a person due to release of radiation from a waste-disposal facility, can be interpreted as saying that one person in a million would die every year from such a cause. Similarly, if X is the concentration of a pollutant in a single sample unit, then the probability that X is less than or equal to x can be interpreted as the long-run proportion of units in repeated sampling which would have concentrations no greater than x.

However, the frequency interpretation is not always suitable for probabilities representing some of the uncertainties which arise in environmental standard-setting. The circumstances of interest may be 'one-off', with no ready representation in relation to a long-term set of events and corresponding frequencies.

Correspondingly, the second standard interpretation of probability is as a *degree of belief*. To say that $P(X \leq x) = \frac{1}{4}$, for instance, is to say that one has the same degree of belief in the occurrence of the outcome of $X \leq x$ as one would have in an event with a frequency probability of $\frac{1}{4}$.

The frequency and degree-of-belief interpretations are not in conflict. Wherever the frequency interpretation holds, it is clear from the above (non-technical) explanation that this probability can also be given the degree-of-belief interpretation. Some situations may be more appropriately represented through one interpretation rather than the other (at least in the view of the person undertaking the study). They also promote different prospects for the analysis of sample data: different approaches to *statistical inference*. See Barnett (1982) for a comparative study of the two interpretations and of the consequences for statistical analysis.

1.5.3   RISK AND RISK ANALYSIS

The term *risk* features widely and is used with a variety of interpretations in many aspects of environmental study (and elsewhere). At one level, it is used as a general term to express uncertainty (synonymous with probability); at another it is used to describe undesirable circumstances which might be encountered and of which we should be aware; risk can also embrace simultaneously both the uncertainty concept and the undesirable prospect. 'Risk analysis' or 'risk management' in this latter context is concerned with a qualitative examination of the portfolio of such risks with a view to assessing the overall situation; 'quantitative risk analysis' augments this with formal (often statistical) methods of analysis and interpretation. 'Risk' is also a technical term in formal decision theory: a different notion altogether.

In view of such ambiguities, we have firmly decided not to express ideas and recommendations in terms of risk-analysis terminology; expression of standards in terms of risk is inherently liable to misinterpretation. Nevertheless, we have sought to ensure that the more important elements of formal risk analysis (as far as these are identifiable) and certainly of decision theory have been incorporated in our discussion.

1.6 SOME EXAMPLES OF STATISTICS AT WORK

We have outlined above the wide range of applications of statistics, integrated into all areas of enquiry and research. A few practical examples may help to consolidate this point.

1.6.1 DIESEL FUEL

The highly multidisciplinary nature of many enquiries is exemplified in a problem of manufacture of diesel fuel for commercial vehicles (Barnett and Lewis, 1967). Below a certain temperature (the cloud point) the fuel jellifies and is unusable, causing serious damage to vehicles. The lower the cloud point of the manufactured fuel, the more expensive it was to produce; major savings could be made for each degree Celsius by which the company could increase the cloud point of its fuel.

The problem was simple to express. Could the company increase the cloud point by 2°C? So many uncertainties were involved in this problem. How did we change the cloud point? How did manufactured cloud point relate to operational problems? What weather conditions would lead to trouble? How did regional location relate to weather and sales? How likely was encountered trouble to lead to brand-switching? What was the overall effect on profitability? Many of these matters were environmental. These questions involved chemists, engineers, sales and marketing staff, psychologists, among others, but central to all aspects of the uncertainty analysis was the statistician.

The crucial needs in this problem were typical. Interaction and intercommunication were of the essence, and the statistical models and methods were complex. Extreme-value analysis and time-series methods were central. There was no difficulty, however, in the statistician becoming immersed with the other professionals, in explaining in understandable terms the statistical implications and in presenting a down-to-earth conclusion readily understandable by all concerned. The statistics were central, complex but benign! With a well-defined level of risk (probability of major difficulties over a 25-year period) the company decided to increase the cloud point by 2°C and made major cost savings over a long period.

1.6.2 WATER DISTRIBUTION AND DISPOSAL

With the privatization of the water companies it became vital for them to assess the current value of stock (piping, pumping stations, sewage works and systems, etc.) and to predict the future costs of maintaining good-quality equipment and service. Current condition was a crucial factor, as were costs of upgrading and development. The quality as well as the supply capability both needed safeguarding – possibly improving! How was this problem to be tackled? Current condition had to be estimated accurately, as did the likely continuing costs and benefits.

Again many specializations were needed in this highly complex problem. Both authors of this report were involved in working as crucial performers in this multidisciplinary problem and the statistical analysis of the uncertainties was central. Costs were extreme: both in analysing the current state of affairs and in developing a monitoring and maintenance policy for the future. The conclusions were bound to be expressed in statistical terms because of all the uncertainties and variabilities. Indeed, this was explicitly demanded by the regulatory agency. One vital problem was how to determine present conditions: to examine even one distribution or sewage location might cost £250 000 or so. Proposals to take a large sample of such observations were hardly likely to be sympathetically received.

Cost-effective statistical approaches involving sample survey methods and Bayesian analysis were developed in cooperation with all the special interests, were explained with care, were accepted and were put into operation. See, for example, O'Hagan and Wells (1993).

1.6.3   GENERAL

Apart from such specific examples, there are whole areas of industry where statistics is now explicitly required and progress depends on appropriate statistical procedures. For example, in the submission of new drugs for approval, or of foodstuffs for production and supply, the regulatory agencies in Britain and abroad require specific statistical evidence of safety and effectiveness. In manufacturing industry, indeed even across the service sector, compliance with quality-assurance and quality-management requirements as expressed in BS 5750 or ISO 9000 is becoming a universal need. These standards necessitate statistical approaches and procedures (for example, methods of statistical process control) at their very heart.

Thus statistics is not only widely applied, but also widely required. These examples are presented to illustrate how successfully (and painlessly) vital statistical considerations were able to be integrated in complex multidisciplinary problems for their efficient and cost-oriented resolution. Communication of statistical approach and outcome was effectively achieved and acted upon. The statistician's role – 'jack of all trades, master of one' (Barnett, 1976) – always involves such interaction and communication and is well honed by widespread and long-term experience in all areas of practical investigation.

1.7   PROBABILITY AND STATISTICS IN STANDARD-SETTING

To review this first chapter, we have shown that environmental pollution standard-setting is a complex process, subject to uncertainty and variation which arise inevitably from various sources at all stages of the process. The proper representation of uncertainty is through probability, interpreted in terms of either long-term relative frequencies or degrees of belief, or both, depending on context.

There is a natural corollary that statistical analysis will also be crucial to the standard-setting process. The assessment and manipulation of probabilities, their use in analysing random data and natural variation, the close interaction with other specialists and careful communication of statistical methods and results are all of the essence of the statistician's profession. Thus statistics must be a vital component in the process of setting standards for environmental pollution. In the following chapters, we shall discuss and illustrate in more detail the treatment of uncertainty and variation in commonly occurring aspects of standard-setting problems. It will become clear that very complex statistical problems can arise, which demand careful treatment.

Basic considerations in setting standards

2

The setting of standards for environmental pollution should be seen as a tool to promote an improved environment and a clearer understanding of how pollutants affect the well-being or health of society or an ecosystem. Section 1.3.1 contained a discussion of various 'kinds of standard'. This chapter analyses the main considerations, and particularly considerations of uncertainty and variation, in deciding what kind of standard to set.

2.1 OUTLINE OF THIS CHAPTER

This is an important chapter which draws together a variety of issues and principles in setting standards. The arguments are sometimes complex, and it is therefore useful to outline the development first.

Section 2.2 is concerned with the *positioning* of a standard. It discusses the important point that a standard may be set at various positions, from the actions which cause pollution at one extreme to the final effects of the pollution on the subject group at the other, and introduces a powerful schematic representation, Figure 2.1, to illustrate this. The question of where to position a standard is one of the most important considerations in standard-setting, and will arise repeatedly in this report. In particular, section 2.2 points out that the costs and benefits resulting from setting a standard are also located at opposite extremes of Figure 2.1; actions needed to meet the standard lead to the cost to the polluter or society of meeting it, while it is through (changes in) the effects on the subject group that the benefits arise.

Sections 2.3 to 2.5 comprise a detailed analysis of ways in which standards are expressed or formulated. Section 2.3 contrasts *ideal standards* with *realizable standards*, and introduces an elaboration of Figure 2.1 in the form of Figure 2.2, in which it is recognized that a realizable standard may be positioned explicitly on the outcome of a sample from the medium or the subject group.

Section 2.4 analyses the effect of uncertainty and variation on ideal and realizable standards. It first makes the crucial point that variation must be acknowledged explicitly in the statement of an ideal standard. This is a point which is often not properly recognized, with the result that some current standards are simply nonsensical. Section 2.4 goes on to examine sample-based realizable standards and the consequences of basing a standard on inherently variable random samples. Again, it is pointed out that some current standards are seriously deficient because of failure to recognize natural variation.

Section 2.5 examines deficiencies in both ideal and realizable standards, and proposes a new formulation – the *statistically verifiable ideal standard*. This is one of the key proposals of this report, which was emphatically endorsed by the Edinburgh Workshop.

Section 2.6 explores the idea of statistical verification in the context of the situation, which commonly applies, where various kinds of standard or objective actually apply at several different positions. For instance, there may be standards on the technology which polluters should use to control pollution, on the levels of emissions to the medium and on the levels of pollution in the medium at locations where it comes into contact with the members of the subject group. The concept of statistical verification leads to a way of defining consistency of standards set in different positions. The idea is explored only very briefly in Section 2.6, and is introduced more as a topic for future study.

Sections 2.7 and 2.8 return to the matter of cost and benefit. Section 2.2 has discussed *where* a standard is set, and sections 2.3 to 2.6 have considered in detail *how* it should be formulated. This final part of the chapter is concerned with *what* level to specify for the standard – that is, the quantitative aspect of standard-setting. We argue in section 2.7 that the level of a standard cannot be set without at least an implicit consideration of the attendant costs and benefits, and that standard-setters should attempt to analyse costs and benefits explicitly, and as carefully as possible.

Section 2.8 examines some phrases used in current standard-setting, such as the precautionary principle. These are seen to be related to other issues discussed in this chapter, particularly to the weighing of costs and benefits. We argue that these may be useful principles to guide the thoughts of standard-setters, but are too imprecise, as presently formulated, to be scientifically acceptable. The precise interpretation and applicability of all these principles can only be seen by careful use of the more general framework specified in this chapter.

2.2   POSITIONING A STANDARD

2.2.1   STANDARDS AS OBJECTIVES

It may be helpful to ask what are the *aims* and *objectives* in setting environmental pollution standards. In general, we think of aims as abstract goals, not necessarily even achievable, which set directions and priorities for an organization. They usually represent policy issues at organizational level. An aim in the context of environmental standards might be, for instance, to prevent damage to rivers, and harm to fish and other aquatic life, from nitrates. Aims will typically be concerned with improved effects on the subject group, and may, like the preceding example, be rather vaguely expressed.

In contrast, objectives should be precise and verifiable. They represent specific achievements which would be clearly indicative of progress towards an aim. We might think of standards as objectives in this sense (or at least as seeking to yield such objectives). In the nitrates example, for instance, it would clearly be a step towards achieving the aim if we could keep nitrates in rivers at levels below those that would harm aquatic life, so we could set this as an objective in the form of a standard for nitrates in rivers.

An alternative which would equally serve as an objective – that is, as an indicator of progress towards the aim – is to control nitrates in drainage from agricultural land. So we could consider trying to set an objective in the form of a standard for nitrates in drains as they feed into rivers.

Another alternative is to tackle the cause explicitly by controlling the use of fertilizers on agricultural land. So another objective might be set by promulgating standards for the types and quantities of fertilizer that farmers can use.

2.2.2   POSITIONING THE STANDARD

Remembering the discussion of section 1.3.2 on balancing cost and benefit, we can see in the above example the possibility of setting a standard at any point on the chain of causality from cost to benefit. Figure 2.1 shows a schematic representation of this chain in a typical application. Subject to the jurisdiction of the body setting the standard, one might set a target, an aim or objective, at any of the four positions shown in the diagram: that is, in consideration of the pollutant effect, the pollutant presence at contact locations, the pollutant presence at entry locations or the actions in relation to the pollutant.

It is important to view Figure 2.1 as representative of a general standard-setting problem only in a schematic sense. We do not mean to

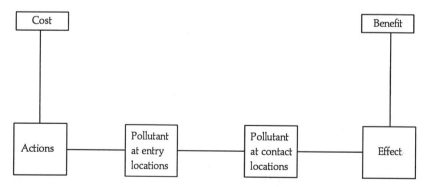

Figure 2.1 Typical chain from cost to benefit

imply that there are only four possible positions for a standard: on the contrary, in practice there are generally many positions along the line from actions to effect at which a standard may be set. Nevertheless, the four generic positions identified in Figure 2.1 will be adequate to illustrate various important points, and this diagram will be referred to at various stages throughout this report.

Nor is it intended to imply that only one standard will be set, in one position, in respect of any pollutant. The matter of multiple standards, and of consistency between standards in different positions, is considered in section 2.6. Furthermore, we argue in section 2.5 that the formulation of even a single standard should often be as a compound of standards which are effectively in two positions.

However, while such niceties are important, it is generally adequate to think of an agency as promulgating a single standard, and Figure 2.1 serves to emphasize that it is necessary to consider where to position the standard.

2.2.3 COSTS AND BENEFITS

In addition to the four locations from actions to effects, Figure 2.1 shows that the costs and benefits arising from setting a standard can be seen as placed at opposite ends of the line. The costs to the polluter (or to society) can be seen as arising from the costs of modifying the actions of the polluter, perhaps by introducing more effective pollution control technology, in order to meet the standard. (There are other costs to consider, as discussed in section 2.7, but these can be ignored for the present discussions.) The benefits, on the other hand, arise from improvements in the effects on the subject group. The relationship of a position to the costs at one end and the benefits at the other end is a significant consideration.

A target for the *effect* (for example, to ensure no harm to river life) achieves a direct benefit, but the necessary practical steps to achieve the target, and consequent costs, are unspecified and need to be deduced through the various steps of the chain of causality. At the other extreme, a target for the *actions* which lead to the pollution (for example, legislation to control fertilizer use) allows costs of implementation to be determined directly, but the consequential benefit is not clear and needs to be deduced through the various steps of the chain.

The intermediate positions, of a target for the pollutant at the points of contact of the subject group with the medium (for example, standards for nitrates in river water generally), and of a target for the pollutant at the points of entry of the pollutant to the medium (standards for nitrates in drains from fields to the river), leave both costs and benefits implicit, to be deduced by working up or down the chain of causality.

2.2.4   UNCERTAIN CAUSALITY

An important point to note is that the causality is not deterministic. As pointed out in section 1.4, there are several sources of uncertainty and variation. Natural variability of the medium (in both time and space) and between individuals of the subject group will combine with scientific uncertainty about the processes involved and with any measurement errors, so that there will be uncertainty either about the changes which would result at any point in the chain from changes at the preceding point or in our assessment of them from sample data. Chapter 3 is concerned with precisely these uncertainties, and the statistical methods appropriate to dealing with them, in its consideration of statistical studies of the pollutant–effect relationship.

Wherever one chooses to position a standard, there will be uncertainty about either the cost or the benefit or both. We shall examine some consequences of this fundamental fact in this chapter.

2.3   IDEAL AND REALIZABLE STANDARDS

2.3.1   POSITIONING IDEAL AND REALIZABLE STANDARDS

In section 2.2.1 we characterized objectives as precise and verifiable indicators of progress towards achievement of aims. We also said that standards could be understood as objectives on occasions, but strictly this implies that standards should be verifiable. Yet a standard which demands, for instance, that the concentration of nitrates in a section of a river should not exceed some limit is not verifiable. We cannot measure the concentration in the whole stretch of river at any one time, still less continuously over a period of time. The standard can be tested by sampling, but this does not objectively verify the standard. Random variation between sample items and measurement errors implies random variation in sample statistics, and a consequent degree of uncertainty over whether or not the standard is being satisfied.

Correspondingly, we will call a standard which is expressed in such a way that one can determine without uncertainty whether it is satisfied at any location a *realizable standard*. A standard which is not realizable will be called an *ideal standard*. In particular, any standard which is set on the level of pollutant throughout a location or over a period of time will typically be ideal because it is not possible to measure the level at all points in a location, or to measure most pollutants at even a single point continuously over time.

We have said that a standard for the level of pollutant at a location cannot be verified objectively by sampling, but if the standard is set only in terms of the sample then it may become realizable. For example, such a standard may specify that nine out of ten samples should have a pollutant concentration below some level, and it is clearly objectively verifiable whether this standard holds. But it is clear that this is no longer a standard for the level of pollutant in the whole medium at that location.

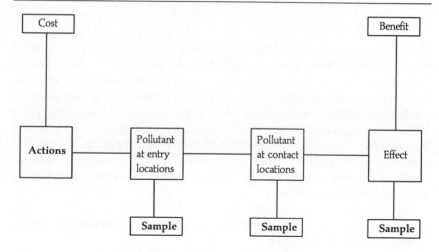

Figure 2.2 The cost–benefit chain with verifiable objectives.

Figure 2.2 augments Figure 2.1 by incorporating the possibility of setting realizable standards by expressing them in terms of sample statistics. We can think of these as further positions at which a standard can be set. Positions labelled in bold face are those at which standards can typically be framed in 'realizable' terms. In addition to the possibility of a realizable standard being based on results of sampling pollutant levels (either at contact locations or at entry locations) or on sampling effects in the subject group, standards on actions causing pollution (for example, a requirement to fit catalytic converters) are typically objectively verifiable and hence constitute realizable standards. Standards on pollutant levels in the whole of the medium at a location, or on effects in the whole subject group, are typically 'ideal'.

An apparent advantage of a realizable standard is that there is no uncertainty over whether it is being complied with. On the other hand, as can be seen in Figure 2.2, placing a standard on a sample will mean more uncertainty over the benefit arising from the standard, because the standard is moved a further step away from the benefit. The corresponding ideal standard will give less uncertainty about benefits, but there will instead be uncertainty over compliance with it. For, while we cannot verify objectively whether an ideal standard is met, it is clear that ideal standards would be completely useless without some operational procedure for determining whether we have 'compliance' with the standard at any location, in some formal sense. We postpone discussion of this question of compliance with ideal standards to section 2.5.

2.3.2   FLEXIBILITY AND COMPLEXITY

Another consideration in the positioning of a standard along the chain from cost to benefit is *flexibility*, together with the attendant consideration of *complexity*.

Consider the question of river-water quality, for example in terms of concentrations of ammonia. The locations at which the same standard is to be applied are stretches of rivers. It would be possible to demand the same standard for every stretch of every river, but it is also possible, and probably much more sensible, to apply standards specific to the characteristics of each stretch of river. The standard might take account of the use of the river for fishing or for drinking-water extraction.

A standard which is to apply at each of a collection of locations can be promulgated in such a way as to take account of the specific circumstances of each location. This gives flexibility but leads to more complex standards. In general it is valuable to take advantage of the flexibility, perhaps to achieve uniformity of benefit, as might be the case with the above example on river water, or to recognize the extra cost of achieving standards in problematical locations. However, more complex standards will inevitably complicate the analysis (which will need to involve at least an element of statistical analysis) of costs and benefits. Consider the effects of air pollution on respiratory illness. The term 'air pollution' is itself highly complex: many polluting agents are involved, each variously affecting different people in different ways in terms of the conditions aggravated by the pollutants, the locations of individuals in relation to the pollutants and so on. The responses are also complex: in terms of outcomes on the various conditions that are affected. Can we even hope to express a joint effect of the pollutant mix on the combination of respiratory conditions? At the very least, such complexity requires a detailed multivariate study of pollution and its effects.

We consider another aspect of flexibility in standards in section 2.5.4.

2.4   ACKNOWLEDGING VARIATION

2.4.1   VARIATION AND THE STATEMENT OF AN IDEAL STANDARD

What does it mean to assert as a standard that the level of particulates in (diesel) engine exhaust fumes should not exceed some value x? To examine this statement, let us first suppose that it is intended that *every* (diesel-engined) motor vehicle should meet the standard (or every new vehicle, perhaps). So the 'location' is a single vehicle and we are setting a standard on 'entry' locations, where the pollutant enters the medium of the air. However, it is still not entirely clear what such a standard would mean because the level of particulates emitted by any vehicle will vary over time. We need to acknowledge the existence of such variation and be able to represent it probabilistically. Yet it is not unusual to see a standard specify that 'the level shall not exceed x' at a location where the pollutant level will inevitably vary from time to time and/or from point to point in the location. This is completely nonsensical. There is no such thing as 'the' level in such a context. Failure to acknowledge natural variation makes for a meaningless and unenforceable standard.

So how should we specify a standard for particulates emissions? We might specify that the level should *never* exceed x, in any circumstances at any time. That would be a very stringent condition. We might specify that the *average* level, averaged over all driving conditions and over time, should not exceed x. That would be much weaker. A vehicle could emit a very high level of particulates some of the time, while balancing that with negligible emissions most of the time.

Setting the standard in terms of the average level would control the total load of particulates emitted by any vehicle over a period of time, and as such might appear to be the most natural formulation. However, we might be concerned about a vehicle emitting very high particulate levels some of the time because that might lead to locally high concentrations in the air, to the detriment of any people who happened to be near the vehicle at the time. Alternatively, it may be the high levels *per se* which have particular effect for some respiratory conditions (such as asthma). So we might need to set a standard to limit both the average (at a baseline level) and the maximum (at a critical threshold). An alternative, when a standard on the maximum is very difficult or costly to achieve, is to say that the emission should not exceed a critical value x more often than a small proportion (such as 5%) of the time.

Similar considerations apply in almost all standard-setting contexts when we wish to specify what we have called an 'ideal' standard. The level X of pollutant at any location will typically vary over time, and will often vary from point to point within the location. For instance, the level of nitrates in a stretch of river will vary both over time and from point to point within that stretch. Any standard for X should recognize this variation. We can think of X as a random quantity with a probability distribution such that $P(X \leq x)$ is the probability that X does not exceed some value x, in the sense that it does not exceed x over this proportion of the time (and/or space). For a single sample unit taken at a random time and/or point, this is represented in terms of the probability distribution of the level of pollutant in that single sample unit.

Figure 2.3 shows what might be the distribution of particulate emissions from a vehicle. The point marked x_A is the average emission (in statistical language, the mean of the probability distribution). The point x_m is the maximum, and formally this value is never exceeded, so that $P(X > x_m) = 0$ (although often in practice a true maximum will not exist: we cannot realistically rule out any values, however high). The point x_p is such that the emission exceeds x_p only a proportion, say 5%, of the time, so that $P(X > x_p) = 0.05$. (In statistical language, x_p is the upper 5% point of the distribution, or the 95th percentile.)

Some ideal standards are expressed in terms which acknowledge uncertainty and variation properly. Others do not quite manage to do so. Consider for instance the EU bathing water directive which specifies concentrations of various pollutants which no more than 5% *of samples* may exceed. It is presumably intended that this should refer to the 95th percentile, which would be the interpretation if '5% of samples' referred to some notional *infinite* number of

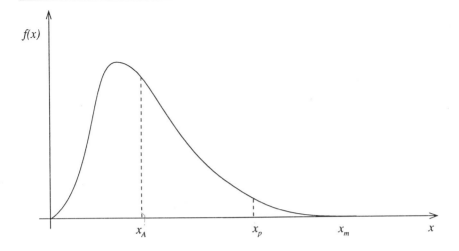

Figure 2.3 A typical distribution of pollutant level.

samples, since then it would effectively refer to the whole distribution of a concentration over time and space. But the statement is hopelessly ambiguous and invites interpretation as a realizable standard based on sample outcomes. (It is of limited value in that case, too, because of not specifying how many samples should be taken – see section 2.5.5.)

2.4.2   CHOICE OF FEATURES TO CONTROL

Any feature of the relevant probability distribution might be made the subject of an ideal standard. We have discussed features such as the average (or mean), the maximum and percentiles. The maximum will generally be a poor choice, first because it will often not be well defined (because no theoretical upper limit exists), but also because it is inherently difficult to control in most applications. A suitable upper percentile is generally a better choice.

The choice of feature(s) to control will depend on context, and in effect on the nature of the pollutant–effect relationship (see Chapter 3). For a pollutant whose effects have a threshold form – negligible effect at concentrations up to the threshold, at which point the effect rises steeply – we will generally wish to specify that some suitably high percentile lies below the threshold, thereby limiting the frequency at which the level goes over the threshold. On the other hand, a standard for a pollutant whose effects are cumulative would more naturally be set to control the average.

Ideally, the choice of feature should be tailored closely to the pollutant–effect relationship. If a level X of pollutant leads on average to an effect $g(X)$, then it seems natural to place the standard on the expectation (a technical statistical term) of the function g. While the relationship will rarely be known so exactly, there may be scope in future standard-setting for exploring this approach.

2.4.3   UNCERTAINTY DUE TO SAMPLING VARIATION

Suppose that an ideal standard specifies that the true level of pollutant X should not exceed a value x_A on average, or that it should exceed a value x_p with probability no more than 0.05. Such a standard cannot usually be verified in practice because we would need to measure the pollutant at all points in the location and/or continuously over time. The only practical approach is to test compliance with the standard by means of a sample. So a sample of some size is taken and a sample statistic (one or more numbers) is calculated from the sample measurements.

From the sample information alone we clearly cannot say definitely whether this location complies with the standard. The use of a sample introduces further uncertainty. Its cause is the same variation in space and/or time of pollutant levels in the medium which necessitates the kind of formulation of ideal standards discussed in section 2.4.1. For instance, if the *sample* average exceeds x_A does this mean that the location fails? Not necessarily. The *true* average of X might well exceed x_A too, but it might be below x_A, in which case it is by chance that the sample average has turned out to be above x_A. Equally, if the sample average is below x_A, this does not necessarily mean that the true average of X is below x_A; or if more or less than 5% of sample units have measurements above x_p then the same does not necessarily hold for the true distribution of X.

The task of determining what can be said about the distribution of X based on a sample is the problem of statistical inference. There are two principal modes of statistical inference. Frequentist inference makes various formalized inference statements based on the frequency interpretation of probability. It would, for instance, declare a level of *significance* for the sample data as evidence for or against the hypothesis that the location complies with the standard. Bayesian inference, on the other hand, makes probabilistic inferences based on the degree of belief interpretation of probability. It would, for instance, declare a *probability* that the location complies with the standard, based on the evidence of the sample. Barnett (1982) provides a thorough comparative study of these principal modes of statistical inference (as well as of some other, minor, modes and variants). For the case of inferring properties of X from a sample, either approach can be used.

Whichever mode of inference is employed, we need to define a criterion, based on a suitable sample statistic, to declare whether a location passes or fails the test of compliance. The important point to recognize is that declaring a location to have passed or failed the test does not prove that it is or is not compliant. Some uncertainty will remain.

The wide variety of contexts of environmental standard-setting problems leads to a corresponding variety of sampling and compliance-testing strategies. The statistical analysis needed to obtain scientifically valid inference statements can be complex. These issues are dealt with in more detail in Chapters 3 and 4 (in relation to pollutant effects, and standards formulation and compliance, respectively).

2.5   STATISTICALLY VERIFIABLE IDEAL STANDARDS

2.5.1   STATISTICAL VERIFICATION

We have said that an ideal standard is not verifiable, whereas a realizable standard is verifiable, in the sense that one can determine without uncertainty whether it is satisfied. For an ideal standard, we have said that it is necessary to augment it with an operational procedure in the sense that we need to take samples in order to test statistically whether it is satisfied. One might argue that this statistical testing is itself a process of verification, and makes an ideal standard verifiable, albeit still with a degree of uncertainty. We could therefore make the distinction more one of contrasting *statistically verifiable (ideal) standards* with *exactly verifiable (realizable) standards*, than of contrasting ideal and realizable standards.

However, there is an important aspect to consider, which may be thought of as relating to the quality of statistical verification. A large and well-planned sample is capable of much more accurate verification than a small or badly planned (or badly executed) sample. The distinction lies in the precision that is achievable in the corresponding statistical inference drawn from collected sample data.

Suppose, for instance, that the ideal standard asserts that the average value of the level X of a pollutant should not exceed a limit x_A. From a sample, we will not be able to claim that the true average has any specific value, but we can, for instance, assert that it lies between some limits x_1 and x_2 with prescribed probability (say 0.95). (These limits are of course calculated from the sample measurements, and technically constitute what are called a *confidence interval* or a *posterior probability interval*.) A poor sample (in terms of its size, design or execution) can produce such limits, but x_1 and x_2 will be much further apart than they would be if calculated from a good sample. With a poor sample, therefore, there is a large chance that when we calculate the limits they will straddle the standard, that is, that $x_1 < x_A < x_2$. Such a result would be inconclusive, since it does not demonstrate that the true average is genuinely below x_A with sufficiently high probability, nor does it demonstrate that the true average is above x_A. A good example will much more reliably produce either clear verification that the standard is met or clear contradiction when it is not.

The formal statistical procedures involved in such inferences can be complex. They will be touched upon further in Chapters 3 and 4, but for our present purposes we need only note that the quality of statistical verification will depend on the quality of the sampling and of the statistical techniques used to process the sample measurements. Essentially, this also involves use of an appropriate *model* to reflect what is truly known about the mechanisms of pollutant effects and about the factors governing uncertainty and variation.

It is our view that an ideal standard should not be set without also setting a standard for its statistical verification. Otherwise, the standard is open to abuse by people claiming compliance on the basis of inadequate statistical (or other) procedures.

2.5.2   THE STATISTICALLY VERIFIABLE IDEAL STANDARD

We therefore introduce the concept of a *statistically verifiable ideal standard*, which we define to comprise two parts.

(a) an ideal standard, which should of course acknowledge variation and uncertainty, as discussed in sections 2.4.1 and 2.4.2;
(b) a standard for statistical verification of the ideal standard, which will typically be expressed as a level of assurance of compliance with the ideal standard that must be demonstrated.

To illustrate how a standard for statistical verification might be formulated, we offer an example based on a statistical testing approach (rather than the 95% interval approach used in section 2.5.1, although the two are not different in important respects). The verification (or test) standard might in this case require that if the true average pollutant level is less than or equal to x_A (that is, the standard is satisfied) then there should be no more than a 5% probability of (erroneously) failing the test. Furthermore, if the true average level equals some value x_B (or worse), where $x_B > x_A$, then the probability of (correctly) failing the test should be at least 90%. Such a standard allows for flexibility in choice of a sampling scheme and statistical analysis, while ensuring quality in the verification process.

2.5.3   EXAMPLE

An example of a current standard which comes close to the form of a statistically verifiable ideal standard is found in the EU's Urban Waste Water Treatment Directive. The standard relates to the concentrations of biochemical oxygen demand, suspended solids and ammonia in discharges from sewage-treatment works. We first express it as if it were in exactly the form of a statistically verifiable ideal standard.

(a) The concentration of x (where x is different for each of the three pollutants) must not be exceeded for more than 5% of the time in any period of 12 months.
(b) On the basis of a number of samples taken over a 12-month period, the lower limit of a 95% confidence interval for the true proportion of time that the concentration x is exceeded should lie below 5%.

This formulation first sets the standard on the 95th percentile, and then bases the statistical verification criterion on a 95% confidence interval. (The two occurrences of 95% are fortuitous and unrelated, but unfortunate because they can cause confusion.)

This is a very interesting example for several reasons, and we shall return to it later, but for the present we should note that actual standard as specified in the EU directive is considerably more prescriptive. In the above form, (b) allows the operator of a treatment works to apply any suitable statistical procedure to obtain a 95% confidence interval with the required property from any suitably collected sample. In fact the standard does not give this flexibility. It actually prescribes a specific statistical procedure (using a particular confidence interval based on the assumption of a binomial distribution). The result is that it specifies how many samples are permitted to have concentrations exceeding x, for a given total number of samples. For instance, if between four and seven samples are taken over the 12 months, then no more than one such sample may have a concentration over x. Whereas if between 41 and 53 samples are taken then up to five may have concentrations exceeding x.

There are advantages in being more prescriptive in this way, one being that statistical verification can be reduced to a simple look-up table for numbers of samples that may be allowed to 'fail'. However, the actual form of the standard does not thereby comply with our definition of a statistically verifiable ideal standard. We return to the question of whether more prescriptive formulations should be offered in section 2.5.6.

2.5.4   CONTRAST WITH IDEAL AND REALIZABLE STANDARDS

The statistically verifiable ideal standard may be viewed as just an ideal standard with a compliance criterion. Just as the ideal standard must be formulated in a way which properly acknowledges natural variation, so the compliance criterion must be statistically based because objective verification is not possible for ideal standards. So a statistically verifiable ideal standard may be thought of as a combination of an ideal standard with a compliance criterion, both of which give proper recognition to uncertainty and variation and are linked to ensure a guaranteed level of assurance that a required condition holds (for example, on the mean pollution level).

Contrast this with ideal and realizable standards. An ideal standard alone is useless. It must be accompanied by a compliance criterion. The kind of verification which can be achieved by any compliance criterion is necessarily statistical, and an example of this is the actual Urban Waste Water Treatment Directive discussed in section 2.5.3. The look-up table of numbers of samples allowed to 'fail' is the compliance criterion. It is an unusually good compliance criterion because the quality of statistical verification which it affords has been thought through carefully. But, as discussed in section 2.5.3, such a standard prescribes a specific compliance criterion, whereas a statistically verifiable ideal standard is more flexible. It specifies the quality of statistical verification required, but does not specify the procedure by which that is to be achieved. This is left open to circumstance, and possibly to negotiation. We believe that this flexibility is an important benefit of the statistically verifiable ideal standard.

In particular, it allows advantage to be taken of improvements over time in technology and statistical techniques, without needing to change the standard. Such improvements might allow the statistical verification criterion to be met more economically, for example by taking fewer samples, whereas an ideal standard with a prescriptive compliance criterion would continue to demand the same sampling effort.

It is clear that a statistically verifiable ideal standard is preferable to any realizable standard based on sampling. This is for two reasons. First, the realizable standard suffers from the same drawback of over-prescriptiveness as an ideal standard with a conventional compliance criterion. Furthermore, a realizable standard based on a sample, as has been remarked previously, is further removed from the benefits in Figure 2.2. The benefits are less certain, and there is even no explicit understanding of how the sampling-based standard controls the pollutant level in the medium as a whole (that is, there is no ideal standard to which it relates).

The only form of realizable standard which might still be usefully applied is a standard set on actions. Such a standard, which is realizable without the need for sampling, and which is therefore not a surrogate for an ideal standard, may be usable and useful. For instance, a standard which demands that all new cars are fitted with catalytic converters is realizable because it can be objectively verified for every car. Statistical verification is therefore not relevant here. In all other contexts, however, we recommend the use of statistically verifiable ideal standards.

2.5.5   STATISTICAL QUALITY AND THE BENEFIT OF THE DOUBT

Some concern was expressed at the Edinburgh Workshop that the introduction of statistically verifiable ideal standards would entail substantially more sampling than is currently done, and hence substantially more cost, because of its emphasis on statistical quality. That is not our intention. The standard should demand an explicit level of quality in the statistical verification process, through the specification of statistical properties that would be required. But the level of quality demanded need not be high. The point is that it should set clear minimum standards for statistical verification, in the same way that the ideal standard sets minimum standards for pollution control, with no overtones intended as to how high the standards should be set. They can be set high or low to reflect costs and priorities.

If we examine our interpretation of the standard for urban waste-water treatment in section 2.5.3 as a statistically verifiable ideal standard, it actually sets a very low requirement. The demand is only that the lower limit of a confidence interval should be below 5%, not that the whole interval should be below 5%. Thus there is no demand that the statistical verification should demonstrate to a required level of confidence that the true proportion of time (that the pollution exceeds the specified concentration x) is below 5%. There is only a kind of double negative requirement, that the sample does *not* demonstrate (at the 95% confidence level) that the true proportion of time is *not* below 5%. This is much weaker and sets a very low standard of statistical verification.

A justification that is advanced for this way of setting the standard is that it gives the benefit of the doubt to the treatment works operator (the polluter). The question of the benefit of the doubt in setting standards is discussed in Lacey *et al.* (1995).

While such considerations are important, and like cost of sampling may lead one to demand a statistical verification that can be achieved by relatively low sampling effort, it can certainly be argued that the standard at present is too weak. For if it is only necessary to fail to demonstrate that the true proportion of time is (with due statistical confidence) above 5%, then this almost invites inadequate sampling. Taken literally, the standard is always met if we have a sample size of zero. Provided no samples are taken, the treatment works always complies with the standard.

We are aware, of course, that this is not how this particular standard is actually applied. The sampling is done by the Environment Agency, which therefore determines *inter alia* how many samples should be taken from each works in a year. The Agency's own guidelines for doing this become in effect part of the standard (but although this gives flexibility, it can be criticized for vagueness and lack of transparency). In fact, perhaps again appealing to benefit-of-the-doubt arguments, the amount of sampling is often very low. The look-up table for allowable failures in sampling admits the case of a sample size as low as 4. Sample sizes as low as 6 are apparently common in practice. Yet if we take six samples and only declare the works failing if two or more samples show a concentration above the critical level x, then there is a good chance of a works which actually gives a concentration above x considerably more that 5% of the time being deemed compliant. For example, if the true proportion of time that the concentration is above x is 25%, then there is only a 47% chance that the works will fail on the basis of six samples. (This calculation is based on statistical theory using the binomial distribution.)

Let us reiterate that it is not our intention to demand high standards of statistical verification, nor to say where the benefit of doubt should be placed in respect of the polluter or the environment. We only insist that there should be a clear standard for the statistical verification of any ideal standard.

It may be that the Agency's intention is indeed to give as much benefit of the doubt to the polluter as the above calculation implies, but the point is that that intention is by no means clear in the standard. We argued in section 2.5.3 that the standard should take the form of an explicitly defined statistically verifiable ideal standard. The form we gave at the start of section 2.5.3 was an attempt to translate the actual standard into a statistically verifiable ideal standard (with advice from statisticians who were involved in framing it). This discussion has shown that even in that form it does not really set a clear standard for the quality of statistical verification. One way to modify our original reformulation so as to set a clear quality requirement is to place a limit on the width of the confidence interval, so that the full standard might then read as follows.

(a) The concentration of x must not be exceeded for more than 5% of the time in any period of 12 months.

(b) On the basis of a number of samples taken over a 12-month period, the lower limit of a 95% confidence interval for the true proportion of time that the concentration x is exceeded should lie below 5%, and the width of the confidence interval should not exceed 20%.

The final figure of 20% (which is purely for illustrative purposes, and not intended as a recommendation) sets the quality standard. If we adopt the same statistical methodology based on the binomial distribution, which was used in creating the look-up table in the actual standard and in our calculation above, then any such limit on the width of the interval implies a lower limit on the number of samples.

2.5.6   IMPLEMENTATION ISSUES

Our proposal of a statistically verifiable ideal standard is new, and we are not aware of any current standards that are expressed in this form. We believe that our proposal merits further study, and are aware that various issues will inevitably arise in its implementation.

The emphasis is on flexibility, on specifying the standard for statistical verification but not saying how that standard should be met, in contrast to the more usual approach of setting a compliance criterion which is prescriptive about how the statistical verification should be achieved. In practice, this may lead to enforcement agencies having to examine diverse statistical techniques which are advanced by polluters in claims that they meet the standard, and having to judge their statistical validity. Polluters, in turn, may be in the position of employing a particular technique but not knowing whether it will be deemed valid by the regulatory agency. Thus where polluters make the case for compliance (but in the scenario where the regulatory agency is responsible for testing compliance) the flexibility may leave them without clear guidance on how to implement the standard and with no independent assessment of the validity of their statistical techniques.

The answer may be for the standard to offer one or more examples of specific sampling and statistical analysis that meets the standard for statistical verification. (The Urban Waste Water Treatment Directive might therefore be phrased primarily as at the end of section 2.5.5, but with the current standard involving a look-up table and a minimal sample size being used as an example.) It may be advisable to set out some probabilistic and distributional assumptions that should or should not be made in any statistical analysis.

The standard might also make demands on the quality of sampling (for instance, requiring an element of randomness or not allowing samples taken too close together in time or space), and perhaps on the quality of measurements made on the samples, that is, on the sampling schemes and principles which would be acceptable (as in the New South Wales Environmental Protection Agency guidelines for sampling of contaminated sites mentioned in section 4.3.1).

These are important considerations, and we strongly recommend that such issues are fully explored in order to prove the applicability, usefulness and soundness of the concept of statistically verifiable ideal standards.

2.6   MULTIPLE STANDARDS

2.6.1   UNCERTAINTY IN THE COST–BENEFIT CHAIN

We have seen that an ideal standard must in general be expressed in a way that recognizes natural variation in the quantity X that is the subject of the standard. We have also seen that in general we cannot be certain that the standard is being met. We now examine the uncertainty that will exist concerning what we achieve in improvement at one position in the cost–benefit chain (Figure 2.1) when we set a standard at another position.

To focus this rather complex argument, suppose that a standard is set to limit the level of nitrates in drainage of agricultural land entering a river. What effect does such a standard have on the level of nitrates in the river? Denote by X the nitrate level in the river, which varies in space (from point to point in the section of river under consideration) and in time, and so X has a highly complex probability distribution. Suppose that we knew exactly what loads of nitrates are entering the river from every land-drainage entry point. Then even with such knowledge we would not know the distribution of X, for several reasons. First, there is natural variation in the flow of the river, which will affect the overall concentration of nitrates. Second, there is variation in the currents which will affect the variability of X from point to point in the river. Among other unpredictable factors are the quantity of nitrates coming from upstream, and the possibility of chemical reactions in the river itself.

In practice, however, we will not even know exactly the (distribution of) levels of nitrates entering the river from all drainage entry locations. We might in fact have samples from these locations but, as we have seen, this means that in various respects we must have uncertainty about the true distributions of nitrates at these entry locations. This compounds the uncertainty about the distribution of X, the level of nitrates in the stretch of river in question.

This example, while not perhaps being an entirely realistic analysis of the problem of nitrates, serves to illustrate a quite general principle, that further uncertainty is introduced at each step in the chain.

Consider the effect in the subject group. Because of natural variation between individuals, the effect will have a distribution. It is this distribution, or more precisely the changes in this distribution induced by improving a standard somewhere, that will determine the benefit of the standard. If we knew exactly what the effect distribution will be, after imposing the standard, then we would know in principle what the benefit will be. But we cannot know the effect distribution precisely even if we set the standard directly on the effect itself.

Now suppose that the standard is set on the pollutant in the medium at contact locations. We might assume, in order to determine its benefit, that the standard is met, but only just. This will not allow us to determine the benefit exactly, because of uncertainty and variation affecting the relationship between pollutant levels at those locations and the effect in individuals of the subject group. So there will be uncertainty about the benefit of such a standard, even if we ignore uncertainty about whether and by what margin it will be met. The uncertainty about benefit is increased as the position of the standard is moved further back along the chain in Figure 2.1. If the standard is set on the pollutant levels in entry locations, then uncertainty about the resulting pollutant levels in the contact locations will compound the uncertainty about the resulting benefit. If we set the standard on the actions causing the pollution, it is actually reasonable to suppose that the standard will be met exactly, but there is now even more uncertainty over what the benefit will then be.

Chapter 3 begins to examine such uncertainties in the links of the chain. Specifically, it takes as its theme the uncertainty in the pollutant–effect relationship, but the same reasoning could be applied to the consideration of uncertainty over any link or sequence of links in the chain. In particular, Figure 2.1 is only intended as a broad-brush representation of the kinds of link that apply in a general standard-setting problem. In practice, there may be many distinct steps in any one of the links shown in Figure 2.1, and other positions at which standards could be set. Each standard-setting problem will have its own unique features in these respects.

2.6.2   CONSISTENCY

Suppose, as will typically be the case, that the principal objective is to achieve some change in the effect distribution among the subject group. This might be set out as a formal ideal standard, and suppose now that it is formulated as recommended in section 2.5, as a statistically verifiable standard. So it includes a requirement for statistical verification to a particular standard.

Suppose, further, that we place a standard on the level of pollutant in either the contact or entry locations. We have discussed in section 2.6.1 how even if we know that this (ideal) standard is met exactly there will still be uncertainty about whether it will influence the effect sufficiently to satisfy the standard set there. This is analogous to the situation of examining some sample data to test for compliance with the effect standard, since that also leaves us with a degree of uncertainty about whether that standard is met. So we can think of the pollutant standard as a test for compliance with the effect standard. Specifically, if the pollutant standard is satisfied, would this be sufficient evidence to verify the effect standard?

By introducing a formal criterion for statistical validation, and thereby creating statistically verifiable standards, it becomes possible to discuss consistency of standards in different locations. Standards are consistent if compliance with one will statistically verify the other. If standards at two positions are inconsistent, then in some sense one is more stringent than the other. This may mean that one is unnecessarily stringent, thereby implying unnecessary costs in those required to comply with it. It may be that one is too weak, so that compliance with it is only one step towards complying with the more stringent standard. Whatever the implication of inconsistency, and whatever the status of the various standards (which might range from legally enforceable to just wishes), it is surely instructive to examine their consistency. This appears to be a new idea that has not been explored in the context of setting environmental pollution standards (or elsewhere). We believe that it merits further research.

2.7   ANALYSING COSTS AND BENEFITS

It is clear that the maintenance of standards involves *costs* (to the regulator, to the complier, to society, etc.) and should *ipso facto* yield *benefits* (to the individual, to society, etc.). It may not be easy to measure the costs or benefits: they may not be easily quantified (especially in monetary terms); they may be multivariate (consisting of several quantities). Nevertheless, standards should be set in overt regard to the input costs and output benefits, at least in principle.

We have devoted considerable space in this chapter to discussing the *form* in which standards are expressed, because we believe it to be important and because we believe that current approaches do not give adequate recognition of uncertainty and variation. There is, of course, the equally important question of how to set the quantitative *level* of a standard. If an ideal standard is to specify that the 95th percentile for the distribution of a pollutant concentration should not exceed x, for instance, how do we choose x? The remainder of this chapter will discuss general principles in setting the level. Briefly, it is our belief that the quantitative levels for standards can only be set by balancing cost against benefit, and that to do so will entail understanding the various links in the chain from cost to benefit shown in Figure 2.1. The modelling of such links, together with the role of uncertainty and variation in understanding them, is the subject of Chapter 3. The remainder of this chapter therefore forms a natural prelude to the next chapter.

In principle, a formal analysis of costs and benefits is a simple application of statistical decision theory (Barnett, 1982, Chapter 7). We consider the possible standards (including the various possible ways of setting the standard as well as the possible formulations and critical levels to use to specify the standard) as a set of possible decisions. In order to compare cost and benefit, both must somehow be measured on the same scale, in the same units. In principle, both are expressed on the scale of *utilities*, and the theory provides a clear definition of the concept of utility which is to be used to convert costs and benefits into utilities. In practice, this entails making exactly the kinds of comparison which make all public decision-making difficult. If the benefit is in lives saved, and the cost in pounds, then the process of converting these to utilities is equivalent to placing a monetary value on a life (surely not an easy task)! It is assumed that a single measure of utility can be set on any combination of decision and outcome. This also is problematical. The simple expedient of defining *overall* utility as the sum of the (positive) benefit utility and its (negative) cost counterpart is not always adequate, and more complex combinations may be called for.

Assuming that utilities can be assigned, the theory allows for uncertainty about the cost and benefit associated with a given decision. That uncertainty implies a distribution of the cost utility and a distribution of the benefit utility and hence a distribution of the conjoint cost–benefit utility. The average (or, in statistical terminology, the 'mean' or 'expectation') of this overall distribution is calculated, and this defines the value of a decision. The optimal decision is the one with the highest value.

Although such a formal decision-theoretic cost–benefit analysis has occasionally been attempted for problems of public policy, most famously in the enquiry into the third London airport, the difficulty and inevitable controversy over the assignment of utilities makes such analyses highly contentious. However, the advocates of such an approach argue as follows. First, any actual decision taken must implicitly have weighed the same imponderables (such as, the monetary value of a life). Second, the decision analysis focuses the discussion on exactly what assignments of utilities must be made, and subject to those determines an optimal decision on objective scientific principles. Third, it is better to make the difficult judgements explicitly and in the open, where they can be the subject of reasoned debate, than to hide them.

It is our belief that such an analysis is the ideal solution to problems of the setting of environmental standards. In practice, we recognize that this can be very difficult, that it will be an enormous task even if feasible, and that it would only be attempted in the most important applications.

Even if a full decision-theoretic cost–benefit analysis is impractical we believe that the level x of a standard can only be defensible if both the costs and benefits that will arise from the level x and from alternative levels have been identified and weighed as carefully as possible. It is therefore always necessary to attempt to understand and analyse the links in the chain from cost to benefit (passing through the position at which the standard is actually set), which is the subject of Chapter 3.

2.8   CURRENT APPROACHES TO SETTING A LEVEL

2.8.1   SOME CURRENT GUIDING PRINCIPLES

Current methodology in setting quantitative standards is very far from a decision-theoretic cost–benefit analysis. Typical approaches include the following.

- 'Safe' levels. Pollutant levels are set which are somehow deemed to be safe. The meaning 'safe'is not often defined, but somehow means that there will be no adverse effect on the subject group, or more realistically that adverse effects will be very rare (acknowledging variation in the effect). Such a standard does not make reference at all to costs, but aspires to maximum benefit regardless of cost. The claim that the level is safe is also not usually justified rigorously with due regard for uncertainty and variation.

- Prudent reduction. A standard for pollutant levels is set at some 'worthwhile' reduction from present levels. This may recognize that a 'safe' level may not be identifiable by virtue of imperfect scientific understanding, or would be too costly to achieve. But again there is rarely any formal analysis to determine either the likely cost or the probable benefits.

- Precautionary principle. This is a broadly applied general principle, if again of a rather informal or subjective nature. In some fields of application it is seen to be 'a driving force behind . . . actions . . . laid down in pollution control legislation' (Scharer (1995), relating to acid deposition on soil). MacGarvin (1995) argues that attempts to determine 'safe' levels of marine biological contamination based on assumptions about the assimilative capacity of the environment 'do not form the basis on which a precautionary principle can be built', thus exhibiting that principle as more than a re-expression of the intent to be 'prudent' or 'safe'. The precautionary principle is linked also with other general prescriptions for dealing with pollution such as the *polluter pays* philosophy (Topfer, 1994).

- BATNEEC. The 'Best Available Technology Not Entailing Excessive Cost' is another approach which recognizes that if a 'safe' level exists it is likely to be too costly to achieve. The standard is set on the actions creating pollution, demanding that an appropriately good technology (BATNEEC) is used to control pollution; see DOE (1991). The standard may alternatively be set on pollutant levels, at levels which are expected to be achievable by BATNEEC, in order to allow the polluter flexibility in adopting a solution. (This allows a choice between equally good technologies and aims at technological advance.) While the cost of such a standard is very clear, and indeed it is set at what is deemed to be a reasonable cost, there is no explicit consideration of what the benefit will be. A related approach, again directed principally at emission levels and included in the Environmental Protection Act 1990 (see DOE, 1996) is the 'best practicable environmental option' (BPEO). Again costs are effectively controlled but effects and benefits are not readily taken into account in any formal or explicit manner.

- ALARA. Another broadly applied general principle for controlling pollution effects operates under the acronym ALARA: 'as low as reasonably achievable'. The key word is 'reasonably'. The intention is that any procedures for controlling levels of pollutant should employ the latest and best technological (and other) aids to achieve outcomes that are ALARA. The implication is that this should ensure safe or prudent levels and that more than this cannot be expected of a body which is generating or controlling a polluting agent. Clearly there are similarities in principle between ALARA and BATNEEC. Indeed, being only informally defined, it is difficult to distinguish between them. The latter refers explicitly to costs; the former surely implies cost considerations in using the term 'reasonably'. The levels declared to be ALARA may be left to the polluter to determine or may be laid down by a regulatory body. For example, Espegren *et al.* (1996) report a study of contamination of properties by uranium products in the context of the US 'Department of Energy ALARA recommendation'. Radiation risk and protection is a major field of application for the ALARA principle, even to tritium contamination from wristwatches (Schonhofer and Pock, 1995) – over 80% of about 30 published papers on ALARA over the past four years were concerned with radiation pollution problems. We might ask how the ALARA principle relates to the use of a formal standard. Sometimes it stands in place of a standard. On other occasions it operates alongside a standard . This is exemplified in Wasserman and Klopper (1993) who, in discussing outpatient radiotherapy by iodine-131 and possible cumulative transfer effects to partners, remark 'in view of ALARA it is recommended that couples sleep apart for at least 14 days after each administration, even if this is below limits permitted by authorities'.

2.8.2 PROPER RECOGNITION OF UNCERTAINTY AND VARIATION

It is our view that while it is inevitable that some pragmatic approach such as 'safe' levels, a 'precautionary principle', BATNEEC or ALARA will be considered in many standard-setting exercises, the final choice of standard must always be justified realistically thorough analysis of costs and benefits (which also must take proper recognition of uncertainty and variation).

If a standard is declared to be a 'safe' level, then 'safe' should be defined, and some scientific analysis offered to justify the effectiveness of the standard in achieving genuinely 'safe' effects. Some realistic assessment of costs is also required if such a standard is to be credible, for if the cost is too high (or if even the very best current technology cannot reach the standard) then it will be clear that the standard is 'pie in the sky'.

If BATNEEC or ALARA is to be used, as well as detailing the cost the proposers should provide a scientific (and possibly statistical) analysis of the consequent benefit. For if the benefit is insufficient, might not the public demand a different interpretation of 'excessive' or 'reasonably'?

This raises the interesting and relevant issue of the distinction between formal scientific evaluation of concepts such as 'benefit' or 'risk' and the public perception of them. These may be vastly different in many respects, from person to person, from time to time, etc., essentially reflecting the different utility functions of different individuals (which in turn include relative facilities to understand and interpret probabilistic and scientific information). Standards have to be 'acceptable' as well as scientifically and statistically appropriate. It is clearly important that issues of 'public perception' are taken account of (in regard to informing, educating and reassuring) in the setting of standards for environmental pollution. See, for example, the recent article by the Chief Medical Officer (Calman, 1996).

Whatever approach is used to settle on a specific standard, some analysis of both cost and benefit should be offered. As we have seen, there will inevitably be uncertainty about either the cost or the benefit, or both, deriving from an assortment of sources of uncertainty and variation. It is therefore inadequate to assert that the cost or benefit 'will be' such-and-such, or even that it is 'estimated to be' something. Either full probability distributions should be given, or else sufficient summaries of those distributions as to convey properly the degree of uncertainty. Echoing the argument of section 2.7 in favour of analysis of costs and benefits, we believe that it is irresponsible to pretend that uncertainty and variation do not exist. A proper scientific assessment of uncertainty and variation is not an expression of weakness but shows honesty and openness.

2.8.3   THE PRECAUTIONARY PRINCIPLE

To illustrate our argument, we sketch here an analysis of the role of the precautionary principle in terms of cost and benefit. The principle is formally a part of European Union law:

> Community Policy on the environment shall aim at a high level of protection taking into account the diversity of situations in the various regions of the Community. It shall be based on the precautionary principle and on the principles that preventative action should be taken, that environmental damage should as a priority be rectified at source, and that the polluter should pay. (Article 130r(2), Treaty of Maastricht)

One definition of the precautionary principle was given at the third North Sea Conference in 1990 (see the DOE UK Guidance Note on the Ministerial Declaration) as follows:

> To take action to avoid potentially damaging impacts of substances that are persistent, toxic and liable to bioaccumulate even where there is no scientific evidence to prove a causal link between emission and effects.

As we have indicated in section 2.8.2, this is extremely vague if taken literally. *Everything* is *potentially* damaging. In the absence of scientific evidence, the principle implies that the only acceptable emission for such substances is zero emission. The only practical way to achieve that is to ban their production and use completely. Surely that was not the intention of those who advanced the principle. It is quite remarkable that what purports to be a scientific principle should be so completely unscientific in form.

Yet underlying this 'motherhood and apple pie' statement is a genuine and important concern which could be explained properly in terms of costs and benefits. In the absence of good scientific information about the relationship between pollutant level and effect, we have to acknowledge that even at very small levels of omission there is a non-trivial probability that a substance which is persistent, toxic and liable to bioaccumulate would produce highly damaging effects. There is therefore a small but non-trivial probability of damage which can be viewed as a large negative cost, if we do not act to control emissions tightly. Therefore, even if the cost of such control is high, the benefit is also potentially very high.

So underlying the principle is the idea that, if scientific information is weak, we cannot rule out the chance of enormous benefits which can balance enormous costs of meeting stringent standards (or enormous damage if we do not act strongly enough). That is perfectly sound in a qualitative way, but we have seen that it can lead to nonsense if taken to extremes. In practice, one would rarely consider the probability or consequences of damage to be so high as to justify an arbitrarily stringent level of control, or banning substances outright. The precautionary principle reminds us that in the absence of good scientific knowledge of the pollutant–effect relationship we must allow for the possibility of great damage from small emissions, but this cannot be turned into a decision on setting a level without properly quantifying the costs and benefits. The principle is not a substitute for proper analysis of cost and benefit, rather a reminder of the importance of such analysis.

We could remark here on the value of scientific knowledge. In general, we have seen that weak knowledge must lead to more stringent standards, on precautionary grounds. The value of science is to allow us, where appropriate, to *relax* the standards.

2.9   CONCLUDING REMARKS

The reader might now wish to reread section 2.1, to review this chapter. Indeed, the chapter contains many important points which will repay careful rereading, to understand fully our proposal in regard to statistically verifiable ideal standards and our recommendations on the analysis of cost and benefit.

The pollutant–effect relationship and other links   3

3.1   LINKS IN THE CHAIN

3.1.1   LINKS IN GENERAL

The chain from cost to benefit was presented in section 2.2 and illustrated in the schematic form of Figure 2.1. It was stressed throughout Chapter 2 that in order to set a standard we must attempt to understand as fully as possible the links in this chain. In particular, if a standard is to be set at one position in the chain, then in order to set an appropriate level for the standard we must try to balance the cost and benefit resulting from any particular level. This means understanding the links (at one end of the chain) between the position at which the standard is set and the cost, and (at the other end) between the standard position and the benefit. Also, if standards are set at several positions then we need to understand the links between the two positions in order to study whether the standards are consistent.

What does it mean to study or understand such links? Consider, for example, the link between actions and the level of pollutant at entry locations. To understand this link we must understand how the pollutant level is affected by actions of the polluter. Of crucial importance is the fact that the relationship between actions and pollutant levels can never be deterministic (in the sense that knowing the actions would mean that we would know the resulting pollutant levels exactly). We can never be free from uncertainty. There will always be uncertainty due to natural variation, even if we do not also have uncertainty about the physical or chemical processes involved. The same is true of any of the links shown in Figure 2.1. If we know the levels of pollutant at entry locations we will not be able to predict exactly the levels at contact locations. If we know the pollutant levels at contact locations we can never know exactly the effects on the subject group. There is uncertainty and variation introduced at every link in the chain. Therefore such links must be described and understood statistically.

We emphasized in section 2.2 that Figure 2.1 is only an aid to general discussion, and that in any real context there are likely to be many more positions at which a standard might be set. Therefore there may be many more links to study. Conceptually, we can always collapse a whole series of links into a single link. For instance, if a standard is set on levels of pollutant in entry locations, then in order to assess benefits it is only necessary to think of a single link between the pollutant level at entry locations and the effects on the subject group; it is not necessary to introduce the position of pollutant level at contact locations which splits the one link into two. Nevertheless, understanding of such *compound* links will generally be obtained most naturally and most effectively by splitting them into several smaller links, and by analysing each simple link separately.

This chapter concerns the processes of modelling such links, and the sources of uncertainty and variation which affect them. Everything in our discussion is intended to apply quite generally to any such link, large or small, simple or compound. However, it is helpful to focus the discussion on links between pollutant levels and effects, since such links are historically the most studied, and are also widely acknowledged (even if only in passing) when setting standards.

3.1.2   POLLUTANT AND EFFECT

Inevitably the relationship between the presence of a pollutant and the effect it has on a vulnerable subject or subject group is riddled with elements of uncertainty and variation. Consider some of these for a water-pollution problem. The pollutant may be a single substance (mercury, say) or a 'cocktail' (of hydrocarbons). Its level at entry locations or at contact locations will vary with spatial position and time. If we seek to measure it there are bound to be measurement inaccuracies. Thus the pollutant *level* X is not a determined fixed quantity but, depending on time or place, it takes one of many possible values. The quantity X may be a single number, or a set of numbers (a *multivariate* pollutant level reflecting different components in the case of a 'cocktail', or different effects of, or surrogates for, pollution in the case of a single pollutant). Even for a specified time or location (which is of limited operational interest) X is not known. It must be measured from sample data, and observed values x will vary because of imprecisions in the measurement process.

What of the *effect* Y? It encounters precisely the same range of possibilities as the pollutant level X. It may be univariate (size of trout population, say) or multivariate (broad diversity measures); it is likely to depend on other factors (including position and time) and thus exhibit intrinsic variation in the subject group; it will have to be measured from sample data, with attendant variation reflecting the relative imprecision of the measurement process.

The presence of intrinsic and random variation of pollutant level and of its effects means that to study either of them requires appropriate statistical methods — even more so if we seek to understand the relationship between the two. It will not suffice to adopt simple summary measures of the data on X

or on Y (for example, sample means or medians, or extreme values). The efficient understanding of the relationship, through the haze of uncertainties and variational elements, requires the use of (often the development of) appropriate sophisticated statistical techniques suited to the complexities of the problem (for multivariate level or effect data, for spatial or temporal variation, to incorporate expert opinion, and so on).

Pollutant–effect relationships in general comprise one area of environmental pollution where an extensive amount of work has been done. Statistical models and methods have been proposed for a wide range of types of pollutant–effect problem; detailed statistical studies have been made of many important pollution issues, and organizations exist to pursue such studies on a continuing basis.

3.2   DESCRIBING AND MEASURING THE POLLUTANT–EFFECT RELATIONSHIP

How are we to describe the relationship between X (pollution level) and Y (its effect)? We must recall how complex each of these can be, and how inevitably complex will be their interrelationship. Their very definitions are problematical. What pollutants and effects are relevant (crucial) in a particular study? How should we express and quantify levels and effects? We must identify the 'key players' and choose the crucial elements of their influence. As remarked above, both X and Y are riddled with uncertainty and variation (in time and space, from subject to subject, due to difficulties of measurement); they are frequently multivariate, rather than each being a single variable.

3.2.1   MECHANISTIC AND STATISTICAL MODELS

A huge variety of forms of model for the relationship between the level X and effect Y is found in practice. We use the term 'model' in a very general sense to describe any of this wide range of formal relationships. We begin by considering two extreme cases.

Sometimes we may believe that we know in scientific terms how the level X of the pollutant relates to its effect Y and we might employ a *deterministic, mechanistic* model which says that Y is some mathematical function $f(X)$ of X.

At the other extreme, we may develop a purely *statistical, empirical* model based on data. We take a set of n measurements, $x_1, x_2, \ldots, x_n$, of X and their associated measured Y-values, $y_1, y_2, \ldots, y_n$. A typical pair will be denoted by (x_i, y_i), where the subscript i ranges from 1 to n. Statistical methods are used to identify, or *fit*, a statistical model from the data. For instance, a regression model postulates that

$$y_i = f(x_i, \beta) + e_i$$

where $f(x_i, \beta)$ is a specified mathematical function but containing an unknown *parameter*, or collection of parameters, β. The term e_i is a *random error* term which recognizes that the observed pairs (x_i, y_i) will not in practice fit exactly the proposed mathematical form $f(x_i, \beta)$ for any value of the unknown parameter(s) β, either because the proposed form is imperfect or because of intrinsic randomness (or, more probably, for both of these reasons). The random errors are assumed to follow some specified form of probability distribution (which may itself contain further unknown parameters). The statistical procedure then estimates the unknown parameters and tests whether the proposed model adequately represents the observed data (x_i, y_i).

Such an empirical model does not purport to express the true, scientific *causal* link between X and Y, merely what form of link seems to be observed in practice, based on whatever data are available. As such it is useful as a purely *ad hoc*, but apparently reliable and adequate, way of predicting the effect Y from the pollutant level X. It can be useful also in suggesting directions in which to look for more scientific understanding.

3.2.2   HYBRID MODELS

In practice, most models will lie between these two extremes. They will combine some mechanistic modelling based on scientific knowledge, statistical elements representing various uncertainties and natural variation, and some empirical modelling derived from observation.

Let us examine how this operates. A purely mechanistic model with no recognition of uncertainty is invariably inadequate because it denies uncertainty which is inevitably present in some form and to some degree. Often a mechanistic model will contain physical parameters whose values are not known precisely. It is common simply to insert estimates of such parameters into the model and so behave as if these estimates were known to be the true values, whereas it is clearly more appropriate to acknowledge explicitly the uncertainty about these parameter values. Furthermore, even if such parameter values are known exactly, one can never really claim that the mechanistic model will precisely predict Y for a given value of X. There is either some uncertainty about the accuracy of the scientific modelling itself, or natural variation present which means that the model can only predict average effect, so that any actual effect will in practice deviate from the average by random error. Typically both forms of uncertainties will apply.

In fact, what we have called a purely mechanistic model is seldom proposed in practice because of its obvious inadequacies. Those who build mechanistic models will generally acknowledge that, because of natural variation at least, their predications can only be of average effect (in some sense). Therefore a full understanding requires a fuller model in which the mechanistic part is augmented by a statistical description of the prediction errors.

This leads to the form

$$Y = f(X, \beta) + E,$$

where β represents the physical parameters, and E represents random error. This makes it look remarkably like the form of the purely empirical statistical regression model presented in section 3.2.1. However, when one begins with a mechanistic modelling approach, the most important difference is that the form of the mathematical function, f, is derived from scientific principles rather than being merely an *ad hoc* proposal. Another difference is that the parameters β are often not estimated statistically from data (x_i, y_i). Each physical parameter is more commonly estimated from a relevant scientific experiment or physical measuring device. Where these are not available, or the evidence is generally inadequate, they may even be estimated by asking experts for their opinions. While mechanistic modellers will usually recognize the need for the random error term E, it is less common to acknowledge the uncertainty in parameter estimates. The model is not complete until these, too, are given proper statistical descriptions.

We have already remarked that, if one starts from an empirical statistical model, the form of the empirically fitted relationship may prompt further scientific study. This in turn will typically lead to modification of the empirical model incorporating some mechanistic elements. Such a model might be termed a *hybrid* model. Most good models become hybrid as one seeks to synthesize observational, scientific and statistical elements.

3.2.3   EXAMPLE: RADIOLOGICAL PROTECTION

We present here an extended example which illustrates how modelling to describe and quantify the pollutant–effect relationship may need to draw together a complex set of components, mechanistic or empirical, deterministic or random. We are indebted to Dr Colin Muirhead of the National Radiological Protection Board for preparing for us the note (Muirhead, 1996) on which this account is based.

The key task in radiological protection is to determine the pollutant–effect relationship between human exposure to radiation (X) and health effects (Y), particularly the incidence of various forms of cancer. The main source of data for analysing such relationships is the Life Span Study (LSS) data from the Radiation Effects Research Foundation in Japan, based on a cohort of 96 000 survivors of the atomic bombs dropped on Hiroshima and Nagasaki, who were followed up between 1950 and 1990 (see Pierce *et al.*, 1996). An estimate of the radiation exposure received by individual survivors is available, together with their subsequent health data.

Modelling of the dose–response relationship for any given cancer from the LSS data will typically involve both empirical and mechanistic elements. The basic relationship may be built on theories of carcinogenesis, but terms to account for age at exposure and time since exposure are often fitted empirically. Use of these models to extrapolate beyond the range of the data is important, and rests heavily on the empirical components. For instance, even 50 years after their exposure, about half of the survivors in the LSS data are still alive, but it is believed that incidence of radiation-induced cancers is heavily dependent on age, even many years after exposure. So extrapolation of an age relationship, to the extremes of the ages found in the LSS and beyond, is important. See, for instance, Muirhead and Darby (1987) and Little (1995).

The LSS data may be the most extensive and useful data set, but it is deficient in important respects. The survivors all received their main radiation exposure in a single event over a short time period, whereas most radiological protection questions concern the effects of chronic, low-level exposure, for example on radiologists or nuclear power-station workers. Evidence for the effects of the rate at which a dose is received comes primarily from animal studies. This is also empirical, and needs to be another component of any realistic model.

It is known also that baseline rates of certain types of cancer vary considerably between Japan and, say, Britain. This adds further uncertainty about extrapolation of analysis of LSS data to other countries. There is almost no good data source, or scientific theory, about the extent to which radiation dose–response relationships will vary between countries.

Other uncertainties arise in the data themselves, because the doses received by the bomb survivors are only estimated and may contain both random and systematic errors (see Pierce *et al.*, 1996).

The estimation of radiation risks is clearly a complex issue, in which careful statistical analysis of diverse but limited data is acknowledged to be crucial.

3.2.4   UNCERTAINTIES IN MODELS

Consider the generic model

$$Y = f(X, \beta) + E$$

presented in section 3.2.2. We now see that we should acknowledge uncertainty in every part of such a model.

- Whether the functional relationship f is derived mechanistically or empirically, we cannot be sure that we have identified the correct form. Further scientific study and more data can both contribute to learning about f, but it will always be an uncertain component of the pollutant–effect relationship.
- Errors of measurement in data, such as in estimates of radiation dose received by atom-bomb survivors, complicate the process of fitting such models statistically. We can think of this as uncertainty in X. Again, when such a model is applied in the process of standard-setting, or in assessing compliance with a standard, there will generally be uncertainty about the value of X (see various discussions of uncertainty about true pollutant levels in Chapter 2). Any observed value Y will also include measurement error and reflect intrinsic variability.

- Parameters β in a model are rarely known exactly, and we have already seen that uncertainty about parameters should be acknowledged.
- The random error E is intrinsically uncertain, of course. Any assumption or model for the probability distribution of random errors (which will often be required) introduces a further level of uncertainty in the form of additional unknown parameters.

(We note in passing that random error has been assumed to be additive here for simplicity of exposition only. Multiplicative or more complex error processes are widely used, too.) We now examine briefly the quantification of these uncertainties, and the process of determining the consequent overall uncertainty about the effect Y arising from a particular pollutant level X.

3.2.5   QUANTIFYING UNCERTAINTY AND ELICITING EXPERT OPINION

Where a model has been fitted statistically, uncertainty about unknown values of parameters within it will be accommodated in the statistical analysis and quantified by means of (for example) estimates and standard errors, confidence intervals or, in a Bayesian analysis for instance, will result in a complete probability distribution for β. However, uncertainty in physical parameters of mechanistic models is not always so easy to quantify. We mention here in particular the case where knowledge about a parameter cannot be obtained from definitive measurements or experiments, in which case one resorts to eliciting the opinions of relevant scientific experts.

The process of eliciting from one or more experts not just a single estimate of a parameter but a full expression of uncertainty about that parameter, in the form of a probability distribution, is fraught with technical difficulties. Psychologists have shown that people can be very poor judges of uncertainties and probabilities, making several kinds of error. Statisticians have studied the process, too, with a view to finding ways to achieve accurate and reliable elicitation. Of three papers presented at a special Royal Statistical Society meeting on elicitation in April 1997, Kadane and Wolfson (1996) give an excellent review of the literature, while one of the examples in O'Hagan (1996) relates specifically to a problem of environmental pollution. See also the comments on ranked-set sampling in section 3.4.1.

3.2.6   WEAK LINKS

There is often some aspect of a model, which can be thought of loosely as a small simple link within the larger complex link that the model is describing, about which very little is known. There may be little or no scientific understanding of the processes involved, and empirical evidence may be scarce. For example, we might try to model the toxicological effects of some substance on humans. Suppose that empirical evidence relates almost exclusively to its toxicology in some other species, such as rats or other primates, and the biochemical mechanisms are not clearly understood. We can think of the relationship between the toxicology in humans and in other species as just such a *weak link* in the model. A similar example is the link in section 3.2.3 between data on Japanese atom-bomb survivors and effects of chronic, low-level exposure in other countries. Weak links may be much bigger: when we face a brand new environmental pollution problem we may view the whole pollutant–response relationship as one big weak link.

We cannot ignore weak links. Our knowledge and understanding of the compound link is broken if we do not somehow model the weak link. But in the absence of hard science or good empirical data we must fall back on the judgements and considered beliefs of those 'experts' who have studied whatever information is available. That is to say, weak links must generally be handled by the processes of elicitation referred to in section 3.2.5.

The usual response to the toxicological example is to apply an extrapolation factor by which to reduce whatever is thought to be a safe dose in rats, etc., to what is deemed to be a safe dose in humans. This is an *ad hoc* adjustment, often itself based on expert judgement. It is not inconsistent with the more thorough approach of eliciting the expert judgements about the link, but it suppresses the uncertainty which is such an important part of the link. In keeping with the theme of this report, we recommend that weak links, like any other aspects of a pollutant–response relationship (or some other larger link of which they form a part), should be studied properly in a way which acknowledges the uncertainty and variation to which they are subject.

Another source of weak links is the use of surrogate measurements, particularly in the context of statistical verification of ideal standards. Ideal standards may be expressed in terms of levels of a pollutant which simply cannot be measured in the required locations, not even at one point within the location and at one instant in time. We have discussed statistical verification in terms of sampling, implicitly suggesting the context where the levels of the pollutant in the samples are used to draw inferences about the level in the whole location. When we cannot actually take such measurements, we rely on surrogate measurements which are somehow related to the measurements we would like to take. The relationship can be thought of as a link which needs to be modelled and understood, and to which all the matters raised in this chapter apply. In particular, when the measurement of interest is so hard to obtain, information about such links tends to be weak.

3.2.7 MODEL VALIDATION

Models should be regarded with suspicion until they are properly validated. The process of building an empirical model, by statistical analysis of data, automatically validates it *on those data*. But statisticians are wary of models which only fit the data used to estimate them, and generally wish to validate their predictions against new data before placing much faith in them. (There is a parallel with the development and confirmation of scientific hypotheses, about which Howson and Urbach, 1996, provide interesting philosophical insights.) The need for validation is correspondingly stronger for models built mechanistically. Faith in the scientific principles on which they are built often lends their developers a touching but misguided faith in their predictions. Such faith is not justified until the models are validated empirically against real data. A cautionary tale is given by Barnett *et al.* (1997) in the context of modelling wheat yield.

Before validation, there must be considerable uncertainty about the correctness or applicability of the functional relationship f that the model expresses. Validation is a statistical process by which that uncertainty is reduced. We should acknowledge that in practice it can sometimes be difficult to obtain data with which to validate models, but this merely emphasizes the need to acknowledge all the elements of uncertainty and variation described in section 3.2.4.

3.2.8 PROBABILISTIC RISK ASSESSMENT AND MONTE CARLO
PROBABILISTIC RISK ASSESSMENT

Probabilistic risk analysis is one name given to the process of assessing the effect resulting from a given level of pollution with full analysis of attendant uncertainties. One methodology for doing this which has become a standard technique is *Monte Carlo* analysis. The essence of the Monte Carlo approach is that random draws are made by computer from the probability distributions of all the uncertain components of a model. Suppose, for simplicity, that only β and E are uncertain in a model of the form $Y = f(X, \beta) + E$, and we wish to use the model to predict Y for a given, fixed X. A random sample of values $\beta_1, \beta_2, \ldots, \beta_N$ is drawn from the probability distribution of β and another sample $E_1, E_2, \ldots, E_N$ from the distribution of E. The size N of these samples is large, thereby ensuring that the samples are good representations of their true distributions. Denoting the generic sample values by β_j and E_j, where the subscript j ranges from 1 to N, we compute a sample $Y_1, Y_2, \ldots, Y_N$ by using the relationship $Y_j = f(X_j, \beta_j) + E_j$. This is then a sample from the distribution of Y, and represents the uncertainty in Y that is induced by uncertainty about β and E. Further detail on the Monte Carlo method can be found in Smith (1993) and Thompson *et al.* (1996).

Monte Carlo is a powerful technique, particularly for models that are not too complex (so that evaluating $f(X_j, \beta_j)$ for a large number N of samples is not itself impractical). It also encounters some criticism, however. See, for example, Moore (1996), although his criticisms really apply to model-based probabilistic risk assessment generally rather than to the Monte Carlo approach specifically. He says that uncertainty about the model f is rarely acknowledged at all in practice, primarily because it is almost impossible to quantify. He points to difficulties in assessing probability distributions for unknown parameters, which we have highlighted in section 3.2.5. He also complains that the results of such an analysis are sometimes misunderstood and mistrusted by decision-makers; we share that concern, but this report reflects our belief that careful statistical analysis of uncertainty is essential, and statisticians should work closely with standard-setting agencies to help dispel that misunderstanding and mistrust.

3.2.9   UNCERTAINTY ANALYSIS AND SENSITIVITY ANALYSIS

An expression with similar meaning to probabilistic risk analysis is *uncertainty analysis*. However, it is generally used in the context of very large mechanistic models. Although in principle it encompasses all sources of uncertainty, in practice it is generally applied to explore the uncertainty in Y induced by uncertainty in input parameters β for the model, ignoring model inadequacy and random error terms. Monte Carlo is again the standard technique, although the fact that the model is large and complex means that a single evaluation of $f(x_j, \beta_j)$ can take substantial computing time. This places a constraint on the size N of sample which can be used, and hence on the reliability of the Monte Carlo method. A substantially more efficient approach, known as Bayesian uncertainty analysis, has recently been developed by O'Hagan and Haylock (1996).

Uncertainty analysis is often preceded by a *sensitivity analysis*, which aims to identify those uncertain input parameters to which the output Y is most sensitive. Uncertainty in other parameters to which Y is relatively insensitive – that is, those parameters for which relatively large changes in value result in only small changes in Y – can be ignored in a subsequent uncertainty analysis. This search may be *ad hoc* or structured, as in a Fourier amplitude sensitivity test (Liepmann and Stephanopoulos, 1985).

3.3   RECENT MAJOR REVIEWS

We now survey the types of pollutant-effect study that have been carried out. It is propitious that two extensive scientific reviews (Barnett, 1997: Piegorsch *et al.*, 1996) have recently been produced on the theme of statistical study of the pollutant–effect relationship. Augmented by the papers presented at a Workshop on Environmental Impact Analysis at the recent Sydney International Statistics Conference 1996 (SISC 96) and the Statistics in Public Resources, Utilities and Care of the Environment (SPRUCE) books (Barnett and Turkman, 1993; 1994; 1997), these sources constitute a valuable digest of the present state of research and data analysis on this theme. They also provide a convenient framework for this report and will be drawn on extensively – particularly Barnett (1997).

Barnett (1997) examines all scientific publications involving statistical analyses of environmental pollution over the period 1991–5 (in particular detail for 1993–5) with the aim of developing a taxonomic base for categorizing the types of pollution issue that have been studied and the various forms of model (statistical and deterministic) and statistical method used in such studies. About 200 publications per year were noted, covering a very wide range of practical problem areas and thus appearing in widely assorted journals and other media.

Only about 10% were in statistical journals, and these were limited to a handful of specific outlets, with particular concentration on the examples of the *Journal of Environmental and Ecological Statistics, Environmetrics* and the publications of SPRUCE. (A new journal of the American Statistical Association, entitled *Journal of Agricultural, Biological and Environmental Statistics,* will extend the scope for specifically statistical work on pollution, although it is to be hoped that such work will continue to be published in applications-based and general statistical outlets.) An extensive range of statistical approaches was evident in this published work, producing an applications-driven breakdown of statistical emphases chiming well with the methods-based categorization of Piegorsch *et al.* (1996), who identifies the following major aspects of modelling and statistical method as relevant to the study of pollution issues, and in particular of the pollutant–effect relationship:

- environmental sampling (with particular reference to monitoring status and trends, to composite and ranked-set sampling, and to adaptive sampling for 'hot spots');
- model selection and model uncertainty (stochastic and deterministic elements, spatial/temporal modelling);
- dose-response and trend testing;
- extreme values and rare events;
- censored data analysis and detection limits;
- combining information from many sources (including hierarchical Bayes methods).

We shall examine in section 3.4 some illustrative examples which reflect many of these directions and methods but which are set in a rather wider classification framework admitting access to a number of identified if not always well-delineated environmental interest themes (discussed in section 3.5) such as methods for 'environmental impact analysis', for critical groups, for critical levels and critical loads, for combining information and for extrapolation.

3.4 SELECTIVE OVERVIEW OF RECENT PUBLISHED WORK ON POLLUTANT–EFFECT RELATIONSHIPS

The publications examined in the preparation of the review by Barnett (1997) covered all pollution media, forms and targets and a wide range of statistical techniques and modelling approaches. We shall survey them in terms of the latter considerations, retaining a catholic degree of coverage of fields of application. (We draw on details of Barnett, 1997, throughout this section and any direct quotations are enclosed by inverted commas and are reproduced by permission of J. Wiley & Sons.)

Of about 500 publications in 1993–5, the predominant emphases yielded the following 'archetypal paper' (amalgamating the predominant emphases in each of the various categories of classification): *work by US scientists published in an environmental science journal using descriptive statistics or simple tests to explore ecosystem effects of inorganic chemical air pollutants.* More specifically, most of the work (45%) was on air pollution, with predominant concern (51%) for gaseous pollutants. Naturally, a major emphasis was on human health issues.

While all the work employed methods (essentially statistical) to represent uncertainty and variation, the majority of studies (67%) did not use anything more sophisticated than simple methods of descriptive statistics or basic tests of significance. About 80% of the work identified the 'review or monitoring' of pollutant effects as its aim. Nevertheless, this still left much scope for detailed modelling approaches, development or refinement of statistical techniques, multivariate, spatial and time-based procedures, new data-sampling approaches, mechanistic models, etc., with applications to many other pollutant media, forms and targets.

3.4.1   SAMPLE DATA

We start with the fundamental question of how to collect sample data for investigating pollutant effects. Sometimes a relevant model for a specific problem will dictate an appropriate data-collection approach. More often, it is important that we observe sensible principles well beyond the notion of 'randomness' to ensure interpretability of the data and their efficient utilization. These ideas should guide the collection of all environmental data (including establishing data networks; cf. the US EMAP programme) and there should perhaps be standards set for the very process of collecting and interpreting pollution data (see section 4.3). A major problem is to ensure comprehensive spatial and temporal coverage to control associated variational factors. While random sampling aims at representativity and interpretability, it may not always suit pollution-effect studies. We often cannot collect data at random. We frequently need (for cost or access reasons) to make maximum use of limited data, and this affects how we should collect the data. Such methods as adaptive sampling, composite sampling, ranked-set sampling and even 'snowballing' may be needed, often with deliberate elements of non-randomness.

Ranked-set sampling is a good example of what can be gained by non-random sampling. Suppose we want to sample pollution levels along a river to estimate the mean pollution level. We might consider taking five or so sites at random to estimate the mean. Alternatively, we might seek advice and information ('expert opinion') to spread out the five observations in an effort to include the worst (highest pollution) prospect and the best (lowest pollution) prospect. The latter type of sample can offer major efficiency gains for the estimation process; over 100% gain in small samples is not unheard of. Ranked-set sampling is a formal statistical method designed to produce and analyse such samples. There is an extensive literature on this topic; see, for example, Kaur *et al.* (1995). Barnett and Moore (1997) provide a recent review, development and application of ranked-set sampling, and offer a ready base for illustrating the ideas and prospects. The method is particularly suited to situations where sampling costs are high but where (prior to actual sampling) it is possible to obtain economic assessments of what are likely to be the relative values of potential observations. In such cases we find that estimating a population mean using ranked-set sampling can turn out to be much more efficient than use of random sampling. The approach goes back 40 years or so. It was first proposed in the context of estimating mean pasture yields.

Ranked-set sampling is particularly useful when costs of analysing samples are high relative to costs of obtaining them. This holds also for *composite sampling*, where we collect many samples (for example, of stream water) and mix them together before analysis. This sounds wasteful; in fact, properly applied, it can be highly cost-effective (see, for example, Edland and van Belle, 1994; and Boswell *et al.*, 1996).

Another departure from randomness is found in *adaptive sampling*, where we start with a random subsample, but allow the results of this to influence where we take subsequent observations. Obviously, pollution is not uniformly distributed. It may 'pocket'; there may be 'hot spots'. Adaptive sampling seeks these out and can lead to efficient estimation both of mean levels and of extremes. Obviously, properly designed statistical methods are required to analyse such 'unrepresentative' data, but these exist. See, for example, Seber and Thompson (1994) for discussion of such an approach for environmental studies; for a particular application, see Englund and Heravi (1993) who 'explore adaptive sampling in the context of soil remediation. They consider a multi-phase sampling approach where the outcomes at each phase are used to determine the sampling plan for the next, applying their ideas to pollution levels at two surrogate sites.'

3.4.2   SPECIFIC APPLICATIONS

Let us pause in our methodological coverage to consider 'some highly detailed and important practical investigations [which] are reported among the pollution/statistics papers in statistical journals. Prominent among these are Gardner (1993), Roberts (1993), Lawson and Williams (1994) and McLachlan (1995). The late Professor Gardner presents a thorough data-based exploration of the problem of childhood leukaemia rates in the vicinity of Sellafield nuclear plant in the UK. Roberts (1993) gives a detailed study of data from sewage plants in Australia with related discussion of what are known as BACI designs ("before, after, comparison, impact").' BACI designs arise quite often in pollution-effect studies particularly where only limited model exploration is possible for access or cost reasons. In this approach, classical experimental design and analysis are not feasible. Instead we seek matched pairs of situations (*vis-à-vis* target individuals or groups, local environment, etc.) and observe each of the members of each pair for some time before introducing a pollutant stimulus to one of them. We then observe each of the members of each pair beyond the intervention stage to see to what extent their subsequent histories differ. Each pair provides a reference set for the effect of a single level of pollutant, and the set of pairs provides data to study pollutant effects over a range of levels. There are difficulties of interpreting the BACI approach, but it can provide information which might not on occasions be available otherwise. 'In a recent special issue of the journal *Environmetrics*, various papers and discussions are addressed to the issue of the public link between aluminium and the expression of dementia in Alzheimer's disease. McLachlan (1995) presents some graphic information . . . [comparing] aluminium concentration in the brain tissue of Alzheimer's disease-affected brains with a control group free of the disease. Lawson and Williams (1994) examine the "Armadale epidemic" of respiratory cancer mortality around a steel foundry in Scotland about 25 years ago using (*inter alia*) a non-stationary Poisson process model, and nonparametric kernel regression.'

3.4.3 SPATIAL AND TEMPORAL VARIATION

'Modelling and analysis of spatial and temporal variation [are] clearly of major importance in the study of pollution. Such work is prominent in the literature, but as in other fields of application manifests rather little progress on the vital matter of *conjoint* spatial/temporal models.

'We find developments of *time series models and methods* representing variation in the time domain (e.g. Carstensen *et al*, 1994, on non-linear time series Kalman filter methods applied to waste-water pollution; and Graf-Jaccottet, 1993, on transformed autoregressive models to study ground [ozone] levels).' 'Time series and Box–Jenkins forecasting feature in the air-pollution work of Gonzalez Manteiga *et al.* (1993). . . . Spatial autocorrelation methods in the context of the growth of industrial pine trees in Georgia (US) are used by Czaplewski *et al.* (1994).' In contrast we find '*spatial clustering* or *dispersal* models for expressing the spatial behaviour *per se* (e.g. Lawson, 1993, who studies directional effects in pollution-induced respiratory mortality near industrial plants; Tirabassi and Rizza, 1994, employing a deterministic advection-diffusion model for air pollution in complex terrain; Biggeri and Marchi, 1995, who consider detection of spatial clusters of rare diseases with a Bayesian extension of earlier methods of Besag, Diggle, etc.; and Gupta and Albanese, 1994, who use survival analysis methods to study spatial contamination of animals by radiation).

'Some work involving both spatial and temporal dimensions is to be found (e.g. Guttorp, *et al.*, 1994, study space-time variation in ground [ozone] levels concluding that separable spatial and temporal correlation structures will not suffice and that a genuinely joint approach is needed!)'.

3.4.4 DOSE-RESPONSE METHODS

It is inevitable that, in studying effects of pollutants, major emphasis should be placed on models and methods for examining dose-response mechanisms. These have been long and widely applied in the medical field and have obvious pollutant-effect carry-over. Essentially, we consider the application (exposure to) successively increasing levels (doses) of a pollutant and changes (increases or possibly decreases) in effects (for example, increasing proportions of destruction of some member of a target habitat). The models and the methods must accommodate the obvious elements of uncertainty and variation, and inevitable nonlinearity in the pollutant–effect relationship. Many such models have been developed and applied (such as, probit and logit analyses). A typical area of application is environmental toxicology. Important distinctions of model and method arise in relation to whether or not successive observations are independent and to whether or not the effect is a count (or proportion) or the value of some continuously variable quantity (such as the level of a biochemical indicator).

Such relational models are examples of the ubiquitous more general regression, linear model and generalised linear models (GLMs), which have importance in pollution studies as much as they do in all other spheres of statistical application. A specific combination of dose-response and GLMs is found in Kerr and Meador (1996) for toxicity bioassay.

Dose-response methods are central to much of the pollution-effect enquiry. It is on the basis of such an approach that principles such as 'no-observed-effect concentration' (NOEC) or 'lowest-observed-effect concentration' (LOEC) are defined to examine standards compliance, and that the newer US Environmental Protection Agency 'benchmark' approach is developed. These are discussed in more detail in Chapter 4.

3.4.5   BAYESIAN METHODS

'A number of valuable papers have appeared adopting a *Bayesian approach*. We have remarked above on the spatial clustering work of Biggeri and Marchi (1995). Bayesian methods are also used in Madruga *et al.* (1994, Bayesian estimates in dose-response studies), in Piegorsch (1994) and in Solow and Gaines (1995). The [latter] two works show an interesting revival in the use of empirical Bayes methods. . . .

'Solow and Gaines (1995) adopt an empirical Bayes approach to the monitoring of water quality.' 'Bayesian decision analysis is the basis of a study to design an interception well to capture a contaminant plume in the work of Wijedasa and Kemblowski (1993).'

Also relevant, although more concerned with monitoring than with pollutant–effect relationships, is the work of Guttorp *et al.* (1993) and Brown *et al.* (1994) on Bayesian interpolation for estimating air pollution.

3.4.6   OTHER STATISTICAL METHODS, AND MODELLING APPROACHES

A wide range of other statistical topics are also in evidence. 'A number of workers adopt a non-parametric approach.

'Apart from Lawson and Williams (1994, mentioned above) we have Korn *et al.* (1994) and Schwartz (1994) concerned, respectively, with the use of combinations of Wilcoxon tests with censored data on well-water contamination by nitrates, and non-parametric smoothing in the study of air pollution and respiratory illness. Schwartz is particularly interesting in the introduction of generalised additive models.'

'*Multivariate methods* (cluster analysis, principal components, pattern recognition) are used by Aruga *et al.* (1993) for river pollution studies (see also Grimalt *et al.* 1993). An interesting application of *"canonical variate analysis"* to examining how fish populations are affected by contaminants is given by Adams *et al.* (1994)'.

Extreme-value methods are most important for examining pollution effects. It is often the extreme values of a pollutant level and its effect that are crucial. ('What maximum level will have no observable effect' – hence, for example, the notion of critical levels and critical loads). Equally, extreme *effect* levels have particular importance in some cases. Methods for modelling and analysing extremes have been in wide use for over 70 years! They play a vital role in hydrology and meteorology, for example, as in determination of 'return periods' for reservoir overflow or severe storms. Applications to pollution effects are to be found, for example, in Smith and Huang (1994) in the context of high threshold exceedances of urban ozone, and in Küchenhoff and Thamerus (1996) for extreme levels of air pollution in Munich.

'*Longitudinal data analyses* and GLMs feature in Burnett and Krewski (1994, air pollution effects on hospital emissions), Corey *et al.* (1994, air pollution and asthma) and McNeney and Petkau (1994, over-dispersed Poisson regression models, again for health effects of air pollution).

'Even *chaos* gets its mention (Paladino, 1994)'; and *neural nets* (Kiilerich and Ruff, 1994)!

3.4.7   DETERMINISTIC (MECHANISTIC) MODELS

Let us recall some aspects of modelling explained in section 3.2. All studies of the effects of pollutants should of course seek to employ what is known about the mechanistic (scientific) links between the pollutant and its effect. There is, however, a wide spectrum of approaches in this regard, from situations where most of the information resides in the empirical link observed purely through data-based experience, through the wide range of stochastic methods where probabilistic models and statistical procedures incorporate the mechanistic knowledge along with characterization of uncertainty and variation, to the opposite extreme where the mechanistic relationship is allowed to predominate and uncertainty and variation are given less regard (possibly only through 'sensitivity studies' of how these latter factors might temper what is essentially modelled as a purely deterministic process: 'data-based mechanistic methods'). As examples of this latter extreme we note the following.

'*Compartmental* models as employed in pharmacokinetics are to be found in the evaluation of toxicokinetic data by Becka *et al.* (1993) and in the study of the health of coke-oven workers by Redmond and Mazumdar (1993).'

Shackley *et al.* (1996) explore the role of mechanistic models for climate change effects, while Young *et al.* (1996) adopt an integrated approach exploring the extent to which proper combination of empirical, variational and mechanistic information is possible (again in climate modelling and in particular regard to the global carbon cycle).

3.5   SOME CURRENT EMPHASES FOR ANALYSING OR CATEGORIZING THE POLLUTANT–EFFECT RELATIONSHIP

3.5.1   ENVIRONMENTAL IMPACT ANALYSIS (EIA)

We have stressed how important it is to assess the 'impact' of an environmental pollutant. This specific approach claims to do just that, but the term is used to cover a range of different types of model and statistical analysis. As remarked above, a workshop was organised under the title 'Environmental Impact Analysis' (joint organiser: V Barnett) at the SISC 96 conference. No specific comprehensive EIA methodology was advanced; different modelling methods (especially spatial/temporal) and different sampling procedures (including BACI; see section 3.4.2) were applied to problems of wildlife, fisheries and radioactive contamination. More general matters of Australian official environmental information systems ('need for a standard for environmental pollution data') and expenditure by industry in the UK in pollution abatement were also covered. Of more direct concern with impact were two papers on, respectively, the inadequacies of the NOEC and LOEC criteria (discussed in Chapter 4) and advantages of the newer effective concentration or EC principle ('concentration associated with a specific level of change in the response relative to control response'), and on the assessment of health effects of inhaled particles and fibres in relation to intensity and duration of exposure.

The need for improvement of 'environmental impact analysis' is discussed by Peterson (1993) in the context of coastal marine studies. The topic also figures in Schroeter *et al.* (1993), Underwood (1993) and Osenberg *et al.* (1994), again featuring BACI sampling designs. Study of the literature suggests that, while interesting topics are discussed, there is no identifiable comprehensive unified theme emerging under the EIA label.

3.5.2   CRITICAL GROUPS

An increasing amount of interest is apparent in approaches to environmental effect and impact based on use of defined subsets of a target population, termed *critical groups*. The idea of the critical group is to seek to identify a subpopulation and a specific subset of pollutants and associated effects which are crucial to an assessment of environmental impact. The critical group is typically intended to be a most vulnerable (or most exposed) subset, often chosen by qualitative (or even subjective) arguments; at least, not on any formally based statistical procedure that takes account of elements of uncertainty or variation.

Recent studies of critical-group applications include Pan (1995) on radiation exposure in the nuclear industry in China; Zeevaert *et al.* (1995) on a model-based ecosystem approach to critical-group methods for waste disposal (see also Zach *et al.*, 1994); Neil (1993) on soil bacteria effects using a compartment model; and MacKenzie and Scott (1993) on Sellafield waste radionuclides.

Wrong assignations of critical-group status can be made. This is evidenced in a comment in the Abstract of Malm *et al.* (1995) on effects of metallic mercury in gold prospecting:

> The goldminers (garimpeiros) that manipulate a major part of the Hg are not the critical group either from exposure to metallic mercury . . . by inhalation or exposure to methylmercury by ingestion of contaminated fish. . . . air and urine sampling show that people working in gold dealers' shops are the critical group concerning . . . [metallic mercury] risk, while riverine communities are the . . . group with respect to methylmercury.

It is clear that the stimulus for the critical-group approach is the existence of an inevitably wide-ranging set of possible effects and impacts due to the presence of various forms of uncertainty and variation. The aim is to 'play safe' by considering an extreme group (why not even an archetypal group?). It is argued that this should help to cope with the variational problems.

On the one hand, the critical group does provide a highly focused basis for studying environmental pollution effects. On the other, it is at best a crude means of seeking to handle variation and uncertainty which could lead to serious misrepresentation of the effects on the community at large if the choice of the critical group proves to be misguided. More seriously (as with judgemental, or even quota, sampling of public opinion) the resulting data are not amenable to statistical analysis or interpretation (for example, to measure final risk) since any subjective element in the process of choice cannot be formally modelled, and there is usually no way of assessing the relationship between the critical group and the population at large (other than in believing it is 'extreme').

It would seem that the critical-group approach should be used with caution: as a first-stage indicator, but not as a basis for detailed conclusion or action. If the aim is to ensure that nobody is exposed to above a certain level of risk in the normal course of life, critical groups could play a useful role, but this would still depend crucially on proper identification of the 'critical' group.

3.5.3 CRITICAL LEVELS AND CRITICAL LOADS

The notions of a *critical level* or a *critical load* (the terms seem sometimes to be used more or less interchangably) are all-pervasive in the literature on the effect of pollution and on the maintenance of environmental pollution standards. The

critical levels or loads are designed to constitute a 'play safe' threshold level
(section 2.9). They are defined in various ways, as the following examples
illustrate:

- 'maximum allowable depositions which do not increase the probability of
  damage to forest soils and surface waters' (Hettelingh *et al.*, 1995);
- 'the amount of atmospheric pollutant that can be deposited on a sensitive
  ecosystem without causing measurable, longterm, degradation in ecosystem
  form or function' (Holdren *et al.*, 1993);
- 'short term and long term average concentration of gaseous pollutants above
  which plants may be damaged' (Cape, 1993);
- 'a quantitative estimate of an exposure to one or more pollutants below which
  significant harmful effects on specified sensitive elements of the environment
  do not occur according to present knowledge' (Strickland *et al.*, 1993).

The basic spirit of these various definitions is the same but there are important
operational differences reflecting inconstancy of the critical-level or critical-load
concept. Thus, uncertainty and variation may not be explicitly entertained
(Holdren *et al.*, 1993) or they may be but in different forms – Hettelingh *et al.*
(1995) use 'probability' while Cape (1993) uses 'average'. The definition of
Strickland *et al.* (1993) seems particularly vague. Measurability is stressed,
however, by Holdren *et al.* (1993) who presumably intend empirical (sample-
based) confirmation but in an unspecified way. As mentioned before, average
values may not be adequate and 'peak values' might cause most damage. Critical
levels and loads are sometimes criticized on such a basis and in other regards
(Ashenden *et al.*, 1996) or praised for their simplicity in providing easily applicable
'instruments for setting effect-related emission standards' (Landner, 1994). Such a
facility is possibly useful in the prime application: namely, the construction of
geographic maps showing vulnerable areas in regard to exceedance of critical
levels or loads for various pollutants and effects. National bodies exist to establish
critical levels, such as, the Critical Loads Advisory Group of the UK DOE.

The critical-load or critical-level concept finds application in many fields of
interest, including effects of air pollution on crops, forests and general vegetation.
The air pollution may take the form of specific agents (ozone) or mixtures of
gases and particulates ('acid rain'). The effects of air pollution on buildings and on
soil, and of air and water pollution on birdlife, are also studied. Soil-pollution
effects on crops and water courses are also subjects of the approach. Sometimes
applications are very specific, such as, critical levels for grazing livestock in respect
of gas-phase and particulate selenium (Haygarth *et al.*, 1994), critical levels of
soil-surface phosphorus in regard to runoff causing eutrophication (Daniel *et al.*,
1993), dry deposition of atmospheric sulphur causing damage to buildings (Butlin
et al., 1995), critical *lower* levels for fresh-water flow into rivers causing damage
to diversity of macroinvertibrate communities (Attrill *et al.*, 1996), and
accumulation of arsenic, cadmium, lead or mercury in the skin of pied flycatchers
(Nyholm, 1995). Sometimes specificity is chosen in a spirit akin to that of

identification of critical groups (section 3.5.2): a specific choice of pollutant and of effect is likely to be the most critical combination needed to be controlled by imposition of an appropriate critical load or level; Nyholm (1995) concludes that 'the Pied Flycatcher appears to be a most applicable bio-indicator for the characterisation of critical levels of pollutants in the atmosphere'.

What are we to make of this approach? It has the advantage of providing simply expressed standards related to the practical situation. It has some regard for uncertainty, variation and sampling, but often in non-specific (non-applicable) form. The definition varies from one situation to another. Its principle is highly conservative (perhaps unduly so). There is no regard for extreme values or actual effect distributions, nor for costs or benefits. See section 4.3.3 for further comment.

3.5.4 COMBINING INFORMATION

Suppose that we are concerned with the relationship between a specific pollutant level X and a particular effect Y but we have access to many sets of data from various sources. These perhaps come from different administrative regions, laboratories, collection agencies, etc., and are likely to be in disparate forms (with different sampling procedures, different variables measured, etc.). This is a common situation, by no means unique to environmental pollution studies. The simple aim is to conduct an analysis of all the data and draw overall inferences from the combined sample information. Achieving this aim is not so simple, however, and has been the subject of much research. A recent compilation volume on the subject is by Draper *et al.* (1992).

Frequently the pollutant–effect relationship will need to be adduced from separate sources of information. It is vital that sound (if complicated) methods for combining information, including meta-analysis, are applied. This is another area where the support of a statistician is crucial.

3.5.5 EXTRAPOLATION FACTORS

The analysis of the pollutant–effect relationship frequently has weak links because of the discrepancy between the real-world context of pollution and the kinds of situation for which data are available. An example is the discrepancy between the good available data on high-burst radiation of Japanese bomb survivors and the context of chronic exposure of European radiation workers, where less precise data are available, a distinction discussed in section 3.2.3. Discrepancy between the real-world and laboratory experiments is almost ubiquitous.

Such discrepancies are rarely amenable to any kind of mechanistic or statistical modelling. The usual response to extrapolating from one context to another is to apply an *ad hoc safety factor*. For example, the Organization for Economic Cooperation and Development (1992) proposes a range of extrapolation factors of 10 000 or 1000 for toxicity of aquatic pollutants, depending on how close the data context is to the real-world context. While such factors are apparently hopelessly crude and unscientific, they may represent the best available estimates of experts concerning the relationship between one context and another. It would be desirable in general to express uncertainty about such factors rather than presenting them as single figures, but we appreciate that this is by no means straightforward (section 3.2.6).

3.6 SUMMARY

In this chapter we have reviewed the crucial topic of understanding and modelling links which are subject to uncertainty and variation. We have exemplified the general case throughout by reference to pollutant–effect relationships. Section 3.1 explains the nature of such links, while in section 3.2 we provide an lengthy discussion of the different ways of modelling and analysing this relationship, taking full account of uncertainty and variation. We then embark (in sections 3.3 and 3.4) on a review of the extensive recently published literature on this topic and conclude (section 3.5) with an examination of less formal approaches to handling the pollutant–effect relationship.

Current and developing incorporation of uncertainty and variability in standard-setting

4

4.1 INTRODUCTION

We have seen in Chapter 3 how extensive is the range of statistical study of the effects of environmental pollution, covering all aspects of pollution and employing almost all fields of statistical methodology.

In broad terms, perhaps 90% of such statistical studies are purely exploratory, examining effects for general scientific or professional interest. Most of the rest is more standards-oriented – prompted by concern for monitoring effects in relation to standards, regulatory requirement or self-imposed constraints, and with examining compliance in such respects. However, there is little evidence of such work being truly linked into, or integral to, the standard-setting process in the sense described above (Chapter 2), or even questioning the fundamental bases on which existing standards have been set.

To some extent this is understandable, since frequent practices of setting standards on informal principles of 'safe levels', 'the precautionary principle', 'prudent reduction', BATNEEC and ALARA (section 2.9), or use of critical groups, critical loads and critical levels (sections 3.5.2 and 3.5.3), provide little of substance on which to explore the appropriateness of the standard, or any formal facility for examining compliance with it. Failure to link standards to observable behaviour or experience further limits the prospects for empirical confirmation.

Placing in perspective the vast literature on statistical studies of environmental pollution, we conclude:

- that the bulk of such work on pollution effects (section 3.4) is very important for what it contributes to our understanding of one of the crucial links in the cost–benefit chain;
- that most of the residue of the work provides an important (if distanced) connection with standard-setting, in respect of its frequent concern for how observed effects relate to currently imposed standards (notwithstanding the fact

that many current standards are of 'ideal' form and thus employ no overt concern for uncertainty, variation and sampling behaviour);

- that a relatively small part of the work yields published proposals for standards which explicitly consider statistical methods to handle uncertainty and variation.

In section 4.2 we shall review some recent publications in the latter two categories: in which statistical studies of pollution are in some way prompted by or related to pollution standards – augmenting the less specifically standards-oriented pollution-effects studies reviewed in section 3.4.

We concluded in Chapter 2 that the most appropriate operational procedure for a pollution standard would involve a conjoint statement of an ideal standard (based as far as possible on objective consideration of costs, effects, uncertainty, variation and benefits) together with a prescription of the extent of statistical verification required to be demonstrated. The wide interest in the literature in the statistical study of sample data relative to existing standards is clearly an implicit recognition of the need for such a sample-based element in standards. The setting of the level for the standard ideally depends on a comprehensive full decision-theoretic cost–benefit analysis. This may seldom be feasible, but choice of level on broader principles – for example, as critical levels (section 3.5.3 and 4.3.4) – must still seek to incorporate as much as is known about relevant costs and benefits.

There are clear indications that interest throughout the world is shifting to the setting of standards which attempt to incorporate properly the multitudinous elements of uncertainty and variation (intrinsic, natural, spatial/temporal, sampling- and measurement-related) implicit in any environmental pollution problem, and away from the extremes of unverifiable ideal standards or uninterpretable realizable standards.

This manifests itself in some evidence of standards changing from *threshold prescription* ('the level shall not exceed . . . ') to operational procedure ('samples must be taken at weekly intervals, for 3 months, over three distinct regions, and the average level should . . . '). This latter style is very much what has prevailed in other well-established and regulated fields, such as, in the pharmaceutics industry or quality assurance (section 1.6), and it will be informative to examine the pharmaceutical procedures briefly in our review (section 4.3) of the changing scene.

We shall note major concern for the statistical handling of uncertainty and variation emerging in the deliberations of leading agencies around the world – in the UK (evidenced in this report to the Royal Commission on Environmental Pollution), in other EU member states, in the USA (in changes under way in the Environmental Protection Agency, or EPA), in Australia and New Zealand (ANZEC, NSWEPA, etc.), and elsewhere. Such developments will obviously need to mature. This will not necessarily be a speedy process. It could be a long time before we encounter as the norm the type of operational standard-setting methods akin to the fully statistically verifiable, or model-based, approaches that we outline in the concluding Chapter 6. This must surely be the ultimate aim, however, and we would urge immediate efforts to begin to construct standards on these principles. We seek to reflect these changing emphases in our report in section 4.3 of personal discussions we have had with individuals in different parts of the world, many of them statisticians in environmental agencies or working on environmental issues. Their interests have ranged over different pollution fields and we have structured our comments in terms of these different interests.

It is important to stress that the views expressed are those of individuals and that they arise in purely illustrative form in respect of topic and emphasis. There is no attempt to provide any comprehensive cover of technical fields or methods. This is true of the whole of this report, as we explained in the Preface, but it is particularly relevant to the present chapter which merely seeks intuitively to convey broad emphases and changes of standpoint both in the literature review (section 4.2) and in the anecdotal commentary of section 4.3.

4.2   STATISTICAL POLLUTION STUDIES LINKED TO STANDARDS INTERESTS

We start by considering some recent publications which discuss in broad terms the statistical effects of environmental pollution in the context of corresponding standards and regulations. These are grouped roughly in terms of the field of application.

Such considerations occur at a general level in the extensive volumes by Hinchee and Olfenbuttel (1991a; 1991b) related respectively to on-site and *in situ* bioreclamation of contaminated land, and in many of the applications publications related to environmental 'incidents' (such as, Chernobyl) or 'issues' (such as, asbestos) referred to in Section 3 of Barnett (1997).

Johnson *et al.* (1995) describes broad issues in the use of data and statistical methods for managing hazardous waste sites.

Warwick and Roberts (1992) quantify by a statistical analysis the risks associated with traditional wasteload allocation analysis, where risk is defined as failing to meet an established water-quality standard.

Ahlfeld and Islam (1994) develop a contaminant transport model for estimating the probability of exceeding ground-water quality standards, and investigate its properties by Monte Carlo simulation.

Work by Guttorp *et al.* (1994) involving spatial and temporal variational components in ground ozone levels is concerned with compliance; more specifically, and in the same practical context, Graf-Jaccottet (1993) uses 'flexible models' of the TBS ('transform both sides') form, augmented with autoregressive measurement-error components, to seek to make decisions in respect of compliance with World Health Organization (WHO) standards (as remarked in section 3.4).

Altshuller and Lefohn (1996) examine reliable estimation of background ozone levels in support of current US EPA re-evaluation of US air-quality standards.

Lee *et al.* (1994) find from statistical studies that the 1979 US standard for ozone is not well related to needs in agricultural production. See also Heuss and Wolff (1993) on ozone control strategies.

Wyzga and Folinsbee (1995) examine health effects of acid aerosols and find evidence of health effects even at levels below current US air-quality standards. Weisel *et al.* (1995) concur, specifically for asthmatics.

Rosenbaum *et al.* (1994) are concerned with statistical mapping of ozone and other air pollutants in the context of agriculture and forestry in the USA. Abatement strategies in the USA are 'standards-based'; the authors argue for an alternative 'effects-based' strategy.

Abraham *et al.* (1993) consider a statistical quality-control system for inter-laboratory comparisons of air-pollution data in relation to German air-quality standards.

Ferguson *et al.* (1995) examine contamination of indoor air by toxic soil vapours in comparison with the current UK Air Quality Standard for benzene (5 ppb).

Solow and Gaines (1995) use empirical Bayes methods for monitoring water quality relative to appropriate standards. Wadsworth and Brown (1995) develop a spatial decision-support system to investigate impact of emissions from major point sources. They minimize 'total environmental impact' in the context of prescribed threshold values.

Reddy *et al.* (1995) use logistic regression models to study compliance with air visibility standards in the US.

Amin and Husain (1994) show statistically that the Kuwait oil fires of March 1991 yielded gaseous pollutants within Meteorology and Environmental Protection Agency (MEPA) standards, but conclude that there were nevertheless prospective human health hazards.

Mikhailov *et al.* (1994) provide regression methods for classifying the effects of climate and air pollution on steel, copper, zinc, etc., in relation to the standard, ISO 9223.

Hipel *et al.* (1995) are concerned with developing a mathematical model involving game theory to ascertain the effectiveness of a reporting system for improving enforcement of environmental laws and regulations when reports are costly.

There are some publications more specifically concerned with employing statistical considerations and methods in establishing relevant standards. Such publications often arise from governmental organizations or public bodies.

The Office of Air Quality Planning and Standards (1996) is concerned with the statistical approach to setting standards for ozone and particulate matter in the USA. It was prepared by the US EPA. This topic will be discussed further in section 4.3.

Correspondingly, DOE (1995) offers a UK statistical standard for particulate matter in the form PM_{10}, expressed as an average over a day. We shall have cause to return to this below (sections 4.3 and 5.3). A parallel Expert Panel on Air Quality Standards (EPAQS) study of ozone is presented in EPAQS (1995).

Ferguson (1992) describes the statistical basis for spatial sampling of contaminated land. See also Ferguson and Denner (1994; 1997) on the derivation of 'guideline values' (or 'trigger values') for soil contamination based on a hybrid stochastic/mechanistic model using a simulation approach (the 'CLEA model').

The use of what is termed the 'benchmark dose approach' for assessing non-cancer health risk is described in Crump et al. (1995) and contrasted with 'no-observed-effect limit' (NOEL) type measures previously employed. These matters will be more fully reported in section 4.3.

A specific statistical contribution to standard-setting is given by Muirhead (1996) who considers standards for radiological protection. He summarizes the data sources available for assessing uncertainty and reviews some advances in developing empirical and mechanistic models.

4.3   THE CHANGING SCENE: ATTEMPTS TO INCORPORATE STATISTICAL ARGUMENTS IN THE SETTING OF STANDARDS

We now turn to the matter of how attitudes to, and emphases in, standard-setting seem to be changing to incorporate an awareness of the centrality of uncertainty and variation. Much of what is reported here has arisen from direct personal contact. It is inevitably anecdotal and selective, but expresses some interesting developments. We are grateful to the many individuals (some identified) for their verbal and written reports and accompanying copies of published material. Space allows only the briefest of summaries.

We have attempted to group the comments under distinct topic heads, but this is not always successful in view of the range of interests covered in some of our discussions:

4.3.1   CONTAMINATED LAND

Much of our discussion inevitably covered issues related to land contamination and hazardous waste.

Dr G M Laslett (DMS, CSIRO, Australia) reports interesting developments in the monitoring of contaminated sites which effectively prescribe what must be achieved for these sites to be judged to satisfy requirements ('standards') of the NSWEPA. Standards Australia has incorporated these procedures in the draft Australian standard on the sampling of potentially contaminated land, but there seem to be some concerns about the statistical propriety of some aspects of the proposals (see the 1994 *NSWEPA Sampling Design Guidelines for Contaminated Sites*) and changes are expected in a forthcoming public review. There is also concern about 'under-sampling' and lack of real knowledge of impact (a researcher was quoted as saying that 'the ANZEC Environmental Investigation Level for lead is 300 mg/kg, but there is no real evidence that 1000 mg/kg in the soil does much harm to human health or the environment').

The NSWEPA report is designed to 'encourage the use of a statistically based approach . . . ' and to 'help and inform environmental consultants, local councils, and others'. The methods have the following form. Soil samples need to be taken both for site characterization and for site validation (the 'standard'). These have to be taken in appropriate numbers, for appropriate depths and sampling patterns, to enable valid statistical interpretation of the results. Specifically, the site will be judged uncontaminated if the data indicate lack of 'high spots' and the 95% upper confidence limit for the sample mean is below a prescribed threshold limit. The procedure admits four types of sampling pattern – judgemental, random, systematic or stratified – and prescribes sample sizes needed for different requirements and different site characteristics. Interestingly, composite sampling (section 3.4.1), is offered as a possible method, but it is pointed out that this may miss 'high spots'. The threshold limit can be chosen either from the existing standard (usually not statistically based) or on 'a site-specific basis using risk analysis'. The latter interesting prospect is not elaborated!

Dr R.J. Lunn (CLUWRR, University of Newcastle, UK) also reports on toxic chemical contamination of soil, in the UK, describing the use of sampling procedures and the interpretation of outcomes relative to declared 'threshold levels'. The threshold levels (whether set by the WHO, the EU or the UK) sometimes appear to be arbitrary (or at least not justified or explained in any formal sense), with no common standard for what sample data characteristic ('percentage') should be compared with the threshold. Sometimes extreme-value distributions and arguments are used (for example, 'the limit should not be exceeded over a period of n years' might be the approach of an industrial company selling land for building). Where data do not exist, or have not been specifically collected, mechanistic (deterministic) models are sometimes used ('validated as far as possible') and simulations are employed to predict ground contamination levels.

It appears that in the USA, the choice of threshold level often figures in the negotiation between a developer and the EPA (or other appropriate agency). So costs will be paramount, even if broader aspects of risk, effect, impact, benefit, etc., are not.

A major report on managing contaminated land is provided by Gallegos *et al.* (1996). The Abstract signals the sophisticated (if non-detailed), nature of the proposals, with shades of risk analysis, decision support, cost-effectiveness, statistics and clear links with compliance if not actual standard-setting.

The ability to make cost effective, timely decisions associated with waste management and environmental remediation problems has been the subject of considerable debate in recent years. . . . a structured, quantitative process is necessary that will facilitate technically defensible decision making in light of both uncertainty and resource constraints. An environmental decision support framework has been developed to provide a logical structure that defines a cost-effective, traceable, and defensible path to closure on decisions regarding compliance and resource allocation. The methodology has been applied effectively to waste disposal problems and is being adapted and implemented in subsurface environmental remediation problems.

The resulting Environmental Decision Support Framework is iterative and cycles through stages of modelling, data collection and analysis, sensitivity analysis and cost analysis. It was prompted by need to implement NRC methodology for low-level radioactive waste disposal. An example is discussed, on disposal of uranium waste in the USA. See also Johnson *et al.* (1995).

An industrial example was outlined by an Australian environmental scientist. An Australian metals company has to monitor sludge for heavy metal and radionuclides before release for soil treatment. Samples are taken from 25 m square areas and stored. For testing compliance, the adopted policy is to accept if the upper limit of a confidence interval is 'well below' the environmental standards for each substance. This approach is not novel; indeed, similar principles seem to operate in the UK, for example, in relation to aspects of water quality. The difficulty resides in the notion of 'well below'. The corresponding cut-off values are often arbitrarily chosen and can lead to varying levels of probability of compliance, lower or greater than that represented by the confidence levels chosen for individual substances. We could in fact tailor the values to achieve a specified overall level of confidence, or of assurance of compliance, which would be a more acceptable approach.

4.3.2 WATER QUALITY

Martin Krogh of the NSWEPA reports an example of a standard (or guideline) for water quality (in terms of presence of faecal coliforms) which is essentially statistical in form. He points out that the Australia and New Zealand Environment Council (ANZEC) guidelines for water quality make a number of broad statistical measurement requirements.

For faecal coliforms the medians of 6 monthly samples are considered and these must be below 150 c.f.u. (coliform units per 100 ml) for the guidelines to be achieved. There are conjoint requirements that more than two individual levels must not be above 600 c.f.u. Other requirements involve less than a 10% change in some parameters (although this idea of change is 'relative' and not well defined). These are strict requirements as the NSWEPA can impose fines or necessary conditions for these guidelines to be met. Note the joint use here of expressions of both 'average' and 'extreme' behaviour. No suggestion was given of a formal joint evaluation of the level of statistical evidence (or assurance) required or provided.

The UK Department of the Environment evidence to the Royal Commission (first two stages) gives a flavour of the extent to which statistical considerations currently figure in setting standards: not entirely absent (for example, radioactive waste, bathing water) but hardly ubiquitous or based on detailed statistical models or methods. In particular, we note a report to the DOE from WRc (Lacey *et al.*, 1995) concerning bathing water standards. The authors give particular attention to the statistical properties of various tests of compliance with UK, EU, US and Canadian standards. They examine the 'quality' of such statistical verification procedures, in the sense of section 2.6. They identify three positions relating to the 'burden of proof' in compliance testing, which give some useful indications of how statistically verifiable standards might be formulated.

It is clear that there is much concern for the proper use of statistical methods in setting and monitoring standards for water quality in the UK. This is further evidenced in the submission from the Environment Agency by Warn (1996) which describes in detail some current approaches and makes recommendations (in the spirit of this report) for future action. We will consider some of these points in more detail in Chapter 5.

4.3.3   AIR QUALITY

There is much interest in and activity on standards for air quality in regard to many types of pollutant and of medium and location.

In an EU Air Pollution Research Report (Ashmore *et al.*, 1990) a review is provided of the effects of air pollution on forests in western Europe. Revolving on the initial sparse knowledge of the link between pollutants and tree development, it is noted that in the 1980s political pressures for reducing emissions of air pollutants were none the less beginning to be translated into action. Such actions were not necessarily based on sound scientific understanding of the causes of forest decline, and of associated potential benefits of the actions, 'but on a "precautionary principle", i.e. that it was prudent to reduce pollutants towards levels at which adverse effect would not be expected'. Some governments took an alternative stance: that it was better to await sound scientific evidence before introducing costly control measures. Since effects were not contained within national boundaries it was felt that international emission control procedures were needed, either on a voluntary basis (as with the so called '30% club') or on a mandatory basis (as with the EC directive on limitation of emissions from large production plants).

The report recognizes the need to act in relation to uncertainties and variation in determining scientifically based, cost-effective control measures on a Europe-wide basis. It refers to 'thousands of papers and reports . . . throughout Europe on . . . forest decline'. A concluding section on implications for future pollution-control policy stresses the importance of models and of the use of critical loads and critical levels. Some emphasis is placed on mechanistic models, but with unknown parameters estimated statistically from observational data, although such an approach is not spelt out in any detail.

Recent preoccupation with critical loads and critical levels of air pollutants as major tools in formulating control policies is noted with encouragement for integration of these ('not particularly well defined') concepts.

In this report, a critical load is informally defined as a 'quantitative estimate of an exposure to one or more pollutants below which significant effects on specified sensitive elements of the environment do not occur'. The notion of critical levels has been used in air-quality standards and guidelines by various organizations (such as, International Union of Forest Organizations). An informal definition is offered in terms of 'the maximum concentration of a pollutant at which adverse effects will not occur on sensitive stages'; this is not very different in principle to that of critical load. Both are expressed in terms of averages over time (for example, an annual mean of 19 ppb SO_2 for the critical level), but it is remarked that the 'averaging times' may be shorter for critical levels than for critical loads. The concepts are not felt to be well defined, nor are the procedures by which they are meant to be applied, but their intent is clear in a general sense and they do seek to make allowance for variational elements. See also German Bundestag (1989; 1990).

Others would claim that the notions of critical level or critical load are unambiguously defined and that they provide natural and important concepts across the range of environmental pollution interests (in the ECEP 19th Report on Sustainable Use of Soil, for example, prominent attention is given to critical loads in relation to controlling emissions of gaseous pollutants into the air and to critical load maps, although the proffered definition is no more precise than those we have considered in section 3.5.3).

R.I. Smith of ITE, Edinburgh, makes a positive case (Smith, 1996), expressing the view that cost–benefit considerations implicitly underlie the use of critical levels and loads by targeting attention to areas of the environment where it is most important to apply available finance. He also sees a clear distinction between critical levels and critical loads: the former referring to 'concentration' and the notion of 'instantaneous' damage, the latter to 'flux' and 'cumulative' build-up of a pollutant (although this may be more difficult to measure).

In general terms, the 'critical level' is sometimes used merely to denote that level which must not be exceeded, that is, as in the statement of an ideal standard or even (if expressed in empirical terms) of a realizable standard.

The US EPA oversees *inter alia* the setting and monitoring of standards for environmental pollution in the USA. Our contact there was Dr Larry Cox, the EPA's senior mathematical statistician, who organized a recent workshop on statistical issues in setting air-quality standards and outlined current activities in these matters.

A workshop was held by the US EPA and National Institute for Statistical Sciences on 18–19 September 1996 to address 'Statistical Issues in Setting Air Quality Standards'. Matters discussed from a statistical standpoint included review of current practices: what statistics are appropriate (e.g. exceedances, spatial averages, etc.) for testing compliance with the aim of protecting human and ecological health, what levels should be set as thresholds (and why) etc. and options for the future in terms of the use of statistics for setting standards and for monitoring compliance with them.

The consideration of methods, policies and procedures for the future is timely, since US EPA is reviewing standards for ozone and particulate matter and entering a 'rule-making' review period in 1997. The present situation and principles guiding possible changes for particulate matter are explained in Office of Air Quality Planning and Standards (1996). A staff paper from the OAQPS reviews the situation for ozone (with over 400 pages of commentary). A critical conclusion is 'that several test statistics should be considered in specifying the term of any new or revised primary standards. Such statistics should include the expected exceedance rate, . . . as well as concentration-based test statistics'. The present standard for ozone is of an ideal type, limiting concentration to 0.12 ppm ozone in the atmosphere, but with a compliance procedure expressed in terms of numbers of exceedances over a 3-year period. The situation is in a state of flux, however, and alternative prospects under consideration for use include such complex and statistically challenging notions as the 17 separate 8-hour running means over a day, and the third largest value of these! (See further details and discussion in section 5.2.2.)

4.3.4   TOXICOLOGY AND DOSE RESPONSE

There is much interest in health effects of pollution, especially reactions and responses of individuals to different levels (doses) of toxic substance.

Dr W Slob of RIVM Holland (the National Institute for Public Health and the Environment) reports (Slob, 1996) on statistical aspects of the assessment of standards for environmental chemicals, distinguishing the determination of 'safe exposure levels' for humans and 'product standards' needing to meet a wider range of environmental objectives. Most work has been based on actual studies, particularly with rodents, using appropriate statistical methods including dose-response analyses, and comparisons with control groups (see, for example, Crump et al., 1977; Crump, 1984) and distinguishing cases where a dose threshold exists below which no effect is evident (non-cancer situations) and where this does not apply (for example, for carcinogens). Results for humans are obtained by 'extrapolation' (presumably to the extent to which this is scientifically justifiable in some sense, although there seems to be limited objective study of such justification).

Let us consider the paper by Crump (1984) in more detail. This is concerned with a new method for determining allowable daily intakes (ADIs) of toxic chemicals by humans (a 'standard' for such exposure). A common method is to determine experimentally (and statistically) a no-observed-effect level (NOEL) (or lowest-observed-effect concentration, LOEC) from animal data and a 'safety factor' (or 'uncertainty factor' as it is commonly termed) to arrive at a level for humans. This approach is also used in the USA to determine threshold limit values (TLVs). Crump (1984) regrets the lack of general guidelines or rules for determining NOELs, the ambiguity of definition of 'no effect' (often depending on subjective judgement), the failure to exploit sound statistical dose-response methodology (or to recognize that the NOEL is only a statistical estimate), ambiguous sample-size effects, and so on. He explains how formal dose-response methods can give better estimates of NOELs for determining ADIs in cases of continuous effect of toxic chemicals, and for quantal ('go–no go') responses. (See section 3.4.4 for further details on various types of dose-response models including logit, probit, generalized linear and other forms.)

Crump concludes that an approach which is better than the usual NOEL one is to fit a statistical toxicological dose-response model to determine (estimate) a benchmark dose (BMD) which represents a 'statistical lower limit on the dose corresponding to a specific increase in risk, over the background level, of between 1% and 10%'. It is suggested that the BMD should replace the NOEL, overcoming most of the perceived shortcomings of the NOEL approach. (Note that again the transference from the animal experiment base to human application is not overtly considered and major uncertainties must remain over the appropriateness of any extrapolation from the animal to the human case.)

It is probably useful at this stage to review the distinctions between the NOEL-type approach and that of the new BMD methodology. Crump *et al.* (1995) describe these distinctions in detail and present an extended study of the variety of dose-response models. Applied to non-cancer risk assessment, the two concepts may be briefly described as follows.

The NOEL (or NOAEL, where A denotes 'adverse') is defined as the highest dose for which no adverse health effects arise. This is not known, of course; it should be estimated from practical data, preferably by fitting an appropriate (validated) dose-response model, although it is often declared 'on the basis of scientific judgement' which can lead to controversy. The NOEL is then translated into a so-called reference dose (RfD) or reference concentration (RfC). Uncertainty enters at all stages. Extrapolation from results of experiments on (often rather few) animals compounds the uncertainties of the form of dose–response curve, from intrinsic variation, from measurement error, etc. The RfD and RfC are further clouded in the attempt to determine them 'by dividing the NOAEL by one or more uncertainty factors'. This approach is all based on a single point on the dose–response curve which needs to be estimated ('statistically or biologically'), interpreted and possibly applied as a standard (effectively of ideal form). The point is intended to be the cut-off between no adverse effect and some adverse effect.

To seek to overcome the many apparent deficiencies of this approach, a new concept of benchmark dose was introduced, directed towards a predetermined level of change in adverse response compared with the response for untreated specimens (usually animals). The BMD is defined as a sample statistic: 'a statistical lower confidence limit on the dose producing the predetermined level of change'. (Note, in contrast, that the NOAEL is a distribution measure, although of course it needs to be estimated). An example of a BMD is a 95% lower confidence limit for the dose producing a 1% increase in adverse response compared with the unexposed (untreated) case.

'The BMD is calculated by fitting a mathematical dose–response model using appropriate statistical procedures'. All elements of uncertainty and variability remain to be accommodated as in the NOAEL approach, although claims are advanced for the benefits of the BMD over the NOAEL ('flexibility', 'accounting more appropriately for sample size and dose–response characteristics', robustness against choice of dose–response model, etc.) Detailed matters of specification, choice of uncertainty levels and dose–response model, use of BMD evaluation, comparison with NOAEL, are considered by Crump *et al.* (1995). The BMD is effectively used in the same way as the NOAEL, for example, in determining RfD and RfC (although their interpretation will differ depending on whether BMD or NOAEL is used). In statistical terms, the difference is between a point estimate of a dose at which an effect is just observable (NOAEL) and a lower confidence limit for a dose yielding a specified change of effect (BMD).

For non-threshold effects, a report by the Health Council of the Netherlands (1994) provides a simple approach to risk assessment used in the Netherlands. In contrast, Moolgavkar *et al.* (1988) advance a sophisticated stochastic model for carcinogens, but this has not yet reached an applications level.

In deriving product standards, these are designed to ensure that only a small percentage of the population (rather than an 'average individual') does not experience total exposure beyond an acceptable level in respect of various different considerations. For a development of this, see Slob (1993), on which basis the Dutch dioxin standard for cows' milk and related dairy products was established.

Mechanistic modelling is again encountered. Toxicokinetic models incorporating statistical components are discussed by Slob and Krajnc (1994).

Ecotoxicological standards to protect whole ecosystems, based on results for single species, have been considered by Aldenberg and Slob (1993).

4.3.5   GENERAL MATTERS

We have remarked on the increasing awareness of, and use of, statistical methods in standards-setting. Some further general indications and conclusions follow.

The US Food and Drug Administration (FDA) has formally announced (*ISI Newsletter*, **20**(2), 1996) the formation of the FDA Statistical Association (FDASA) to promote the advancement of statistical sciences within the regulatory environment of the FDA and to foster FDA-wide consistency and harmonisation on crucial regulatory statistical issues. The FDASA has a current membership of approximately 100 professionals. (Information: Dr Margaret Lamb, Dept. of Health and Human Services, FDA, Rochville, MD 20857, USA).

The US *EESI Weekly Bulletin* of 29 July 1996 reports that 'House, Senate pass comprehensive pesticide reform'. The new bill sets a 'strong health-based standard, provides special safeguards for children and protects . . . health and safety using the best science available', according to President Clinton. Statistical considerations underlie this new procedure. The EPA is required to set a uniform pesticide tolerance for raw and processed foods that is 'safe', that is, provides 'reasonable certainty of no harm' interpreted to be a 'one-in-a-million lifetime cancer risk'.

What are we to conclude from this assortment of views and developments? We might draw the following tentative conclusions.

It would seem that in the early stages of standard-setting there was little by way of relevant data and no extensive use made of statistical methods to handle uncertainty and variation. Examples that came soonest and closest to statistically based standards arose in relation to toxic substances, where the standard itself seems to require a statistical procedure to be carried out to assess if levels exceed a critical threshold value (but this value is arrived at by *ad hoc* adjustment of levels determined from laboratory experiments on animals); and in air pollution, where the compliance condition for the ozone standard requires regular readings for ozone to be taken hourly over a spatial grid of typically 10–12 (or fewer) monitors in a geographic region and where if more than three exceedances occur in 3 years the region is regarded as non-complying or in violation. See Chapter 5 for more details.

At the current time we can readily identify three cases, in relation to the level of a standard and monitoring its compliance. There is little evidence of statistically verifiable ideal standards, or of detailed cost–benefit methods employed to choose the reference level of the standard. Instead we distinguish cases of:

(a) an *ad hoc* standard level and no statistical rules for compliance;
(b) an *ad hoc* standard level and statistical rules for compliance;
(c) a statistically evaluated standard level and statistical rules for compliance.

It would seem that at present and throughout the world most standards fall into category (a) some into (b) and few, if any, into (c).

4.4 PROGRESSION

In this chapter we have noted (section 4.2) a body of published work using statistical methods, which goes beyond the pollutant–effect relationship and is related to standards and regulations *per se*. We have also considered, albeit in a self-selective and anecdotal manner (in section 4.3), some expressed views on the use of statistics in standard-setting (where procedures are being adopted or developed which formally incorporate methods to handle uncertainty and variation) and some indications of changing attitudes and emphases which are supporting (in fact demanding) the use of statistics in standard-setting. So far in this report, this requirement for the use of statistical methods has been argued and illustrated at various levels, namely in regard to:

- the fundamental roles of uncertainty and variability in the standards context (Chapter 1);
- principles for setting standards (Chapter 2);
- effects of environmental pollutants (Chapters 1 and 3);
- standards monitoring and compliance (Chapter 4).

It is appropriate for us now to stand back and attempt to give a broad evaluation of present trends and a proposed formal (model-based) principle for standard-setting in the future, always in the firm conviction that uncertainty and variation must be incorporated. We deal with these matters in the concluding Chapter 6, after (in Chapter 5) discussing three specific examples (or sets of examples) of standard-setting in practice: the first involving fairly sophisticated use of statistical methods for handling uncertainty and variation; the second acknowledging and attempting to meet such a need but with only limited success; and the third showing no real concern for such matters.

Current standards: Examples

5

5.1 THE EXTENT TO WHICH UNCERTAINTY AND VARIATION ARE CONSIDERED

Standards and regulations governing environmental pollution have been in operation for a long time. They are continually being revised as we acquire more knowledge of effects, costs and benefits. We are also forever refining our knowledge of the effects of uncertainty and variation on the behavioural characteristics of the various environmental pollutants. It is natural to ask to what extent this improved knowledge is influencing the changing standards. More specifically, we might look for examples of standards actually being set with the incorporation of the many facets of uncertainty and variation and procedures for handling them along the lines we have explained in this report. In section 5.2, we take a look at some specific standard-setting examples. Specifically we examine whether, and to what extent,

(a) account is taken of uncertainty in links such as the pollutant–effect relationship, as discussed in Chapter 3, in selecting a suitable level for the standard;
(b) recognition is made of inherent variation and uncertainty when expressing the standard, as discussed in section 2.4.1;
(c) a suitable verification procedure is defined for an ideal standard, with corresponding assurance of the quality of statistical verification, yielding what amounts to a statistically verifiable standard, as discussed in section 2.5.

In section 5.2 we present three environmental pollution standards which come closer to fulfilling these three aspects of performance than any other examples we are aware of. Each acknowledges the importance of uncertainty and variation but none meets all the conditions, and we are not aware of a standard currently applied which does so. The first example, on the quality of river water, exhibits an ideal standard with a clear, statistically based compliance criterion and a method of setting the level which pays regard to costs and benefits. We have detailed concerns about the formulation of the standard and the treatment of costs and benefits, but this standard seriously attempts to address all three of the above points. The second is the current US standard for ozone, which consists of an ideal standard and a compliance procedure, but offers no justificatory basis for choice of level in the ideal standard, is prescriptive in the form of the compliance procedure and does not consider the level of assurance provided. Our third example, on disposal of low- and intermediate-level nuclear waste, places commendable effort on modelling and analysing action–pollutant–effect links and benefit taking the form of a standard defined in terms of effect, but costs are not objectively examined and no verification procedure is proposed.

Section 5.3 considers examples of standards which have little or no regard whatsoever for any uncertainty and variational factors, and offer no basis for assessing compliance either by the 'prospective offender' or by the 'regulator'. Unfortunately, such standards are not uncommon.

5.2 EXAMPLES WHERE QUITE SOPHISTICATED TREATMENT OF UNCERTAINTY AND VARIATION IS EMPLOYED

5.2.1 STATUTORY WATER QUALITY OBJECTIVES

Following EC directives, it is intended to set statutory water quality objectives (SWQOs) for all but the smallest stretches of river (and, where appropriate, canal) in the UK. Responsibility for proposing and monitoring SWQOs lies with the Environment Agency (EA) which now embraces the old National Rivers Authority (NRA). (In Scotland, the relevant agency is the Scottish Environment Protection Agency, but we shall not complicate the presentation by making this distinction explicitly hereafter.) An SWQO will place a stretch of river in one of five categories, RE1 (the highest standard) to RE5 (the lowest). For instance, in RE1 the SWQO will demand that the river contains 80% or higher saturation of dissolved oxygen at least 90% of the time, whereas in RE4 the target 80% saturation is reduced to 50% and in RE5 to 20%. We will first examine the RE1 to RE5 standards in detail with regard to their proper acknowledgement of uncertainty and variation, and their inclusion of suitable statistical verification requirements.

The standards are expressed as ideal standards, and generally give recognition to uncertainty and variation. Each contains requirements on dissolved oxygen, biochemical oxygen demand, total ammonia, un-ionized ammonia, pH, dissolved copper and total zinc. All are expressed as percentiles, thereby acknowledging variation over time and over spatial position in the river stretch. There is, however, ambiguity concerning the ranges of time and space which define the probability distributions to which the percentiles are to be applied. In NRA (1994), the EA's interpretation is set out: it defines the time range to be the hours of normal working over a year (thereby excluding the night, weekends and holidays), and will typically work with only a single point (chosen to be representative of the part of the river stretch with stable conditions) in the stretch.

Compliance with the ideal standard is assessed by a procedure very similar to that described in section 2.5.3 and discussed at various places in section 2.5. It differs by using a more sophisticated technique based on assuming normal or lognormal distributions for sample pollutant measures, rather than simply on whether each sample measure exceeds the level in the ideal standard, but otherwise the comments we made in Section 2.5 about the Urban Waste Water Treatment standard apply. In short, the compliance criterion is statistically sound, but it is prescriptive rather than taking the form of a statistically verifiable ideal standard, and gives substantial benefit of the doubt to the polluter without demanding minimal quality of statistical verification.

We turn now to the matter of setting the level of the standard, and the extent to which the level is set with due regard to, and with proper understanding of, the costs and benefits that would ensue. The five river ecosystem standards RE1 to RE5 provide a range of levels among which to choose when setting the SWQO for a river stretch. The EA proposes a standard, and thereby chooses a level, with quite explicit consideration of cost and benefit. It has to date prepared proposals for eight catchments in England and Wales. These proposals are set out in consultation documents which describe the proposed SWQOs for stretches within the catchment, as well as costs and benefits. For example, the proposals for the Worcestershire Stour Catchment (NRA, 1996) describe benefits for each stretch in terms of potable supplies, agricultural and industrial abstraction, fisheries, river ecosystem, recreation, amenity and aesthetics, and tourism, classifying the benefit in each category for each stretch as high, medium or low. The costs are set out in terms of actions which have been identified for polluters, and in most cases costed. For instance, in the whole of the Worcestershire Stour Catchment, comprising 164 km of watercourses divided into 48 stretches, improvement work needed by Severn Trent Water Limited to meet the proposed short-term SWQOs is costed at over £55 million. Although costs and benefits are not weighed in these proposals by any sort of decision-theoretic analysis, the document has been put out to consultation so that a broad range of views on the value of the benefits in comparison with the costs may arise.

The relationship between proposed SWQO standards and the identified costs, in particular, rests on some detailed modelling. Both Monte Carlo simulation and mathematical analysis (the 'Warn–Brew method') are used to model the relationship between discharge consents for water treatment works and industrial effluents, on the one hand, and pollutant levels throughout the catchment on the other hand. Uncertainty is handled through further simulation-based uncertainty analysis using a program called SIMCAT. While these models and methods may be open to various improvements, they have been validated in a range of contexts and represent a good attempt at analysis of the relationship between actions and pollutant levels. The relationship between those levels, RE1 to RE5, and the benefits is less well established, but the implications of the five levels for river ecosystems (including fish habitats, water birds, etc.) and general amenity are broadly understood (and this understanding guided the setting of these five levels by the EU).

In conclusion, we can summarize our analysis in terms of the three criteria set out in Section 5.1 as follows.

(a) There is a serious attempt to understand both costs and benefit and the uncertainties attaching to both. In particular, the identification of costs, through not only identifying required actions but also a clear analysis of the relationship to provide good statistical assurance that those actions will be adequate, is among the best analyses that we know of. Although the balance between costs and benefits is not set through a scientific decision-theoretic analysis, careful consideration is facilitated by a very clear consultative document.

(b) The ideal standards are set with proper recognition of uncertainty and variation. The only deficiency in this regard is some ambiguity over the range of space and time to which they apply.

(c) The statistical verification procedure adopted by the EA is sound but prescriptive, and the quality of verification is not properly assured. Nevertheless, only minor modification would be needed to make this a genuine statistically verifiable ideal standard.

5.2.2   THE US OZONE STANDARD

The US EPA is responsible for air-quality standards. The present primary and secondary standards for ozone were introduced in 1978, revising earlier standards of 1971 which applied to several airborne components, not just ozone. The 1978 primary and secondary standards were set at 0.120 ppm and (see Office of Air Quality Planning and Standards, 1996) claimed the novel feature of being 'revised from a deterministic form to a statistical form'. That is to say, the standards purported to take cognizance of uncertainty and variation. Specifically, the standard was expressed as follows: that 'the expected number of days per calendar year with hourly average concentrations greater than 0.120 ppm is equal to or less than one'. The standard is thus expressed in terms of the mean of a distribution, namely the distribution of the number of days in the year when the maximum value of the mean hourly ozone level is greater than 0.120 ppm. Short-term temporal variation is taken into account by using the daily maximum of the hourly average, and we are then incorporating seasonal variation by taking an average over the year. With regard to spatial variation, the standard is expected to apply everywhere (that is to say, at any designated position).

So probabilistic considerations are allowed to play an important role in the specification of the standard *per se*. (The 1971 standard was not of this form; for example, the primary standard had the 'deterministic' definition of 'an hourly average of 0.08 [ppm] . . . not to be exceeded more than one hour per year', although this in fact still embodies a probabilistic notion of variation from day to day.) But the problem is that the standard in its basic form is a statement about the true ozone level everywhere. As expressed, it provides no procedure for self-monitoring or for examining compliance. It remains an unverifiable ideal standard. Clearly this was recognized, since a compliance procedure was introduced to be operated in terms of performance over a network of sites introduced by EPA to monitor ozone levels (and levels of other air pollutants). This operates as follows. An agency is in violation of the standard if the number of occasions (days) on which the hourly maximum ozone level exceeds 0.120 ppm in a 3-year period is more than 3.

The standard and the compliance procedure thus seem jointly to constitute what is effectively a statistically verifiable ideal standard. Or do they? This issue has been clarified in helpful discussion with Peter Guttorp of the University of Washington in Seattle. He has provided access to US EPA documentation of 1979 which, in amending sections 109 and 301 of the US Clean Air Act, makes it clear that from that stage the compliance procedure is attached to the statement of the standard in the following terms: 'The standard is attained when the expected number of days per calendar year with maximum hourly average concentration above 0.12 ppm . . . is equal to or less than 1 as determined by appendix H'. Appendix H then explains that, in effect we observe on how many days over a 3-year period the daily hourly maximum concentration exceeds 0.12 ppm, and if this is no more than three, the area is in compliance with the standard.

But is this conjoint standard/compliance criterion really what we have called a statistically verifiable ideal standard? There are matters which mark it out not to be of this form. In the first place, the compliance procedure expresses what constitutes violation. However, it is not formulated in terms of a specified probability that the standard is itself violated. It is also totally prescriptive and leaves no freedom for the complier to choose how to demonstrate any specified level of assurance.

Basically, we need a standard which requires the complier to demonstrate with prescribed probability that the level specified in the standard is not being breached. While not strictly expressed in this form, and with no procedural latitude, perhaps the 1978 ozone standard nevertheless achieves such an objective. It would do so if, whenever the standard is being breached, the implied probability that this would be picked up by the compliance procedure is at an acceptable level.

We can calculate this probability. Suppose we are on the margin of breaching the level and on average there is precisely one day per annum when the mean hourly maximum level exceeds 0.120 ppm. This is equivalent to saying

$$P(\text{max level in day} > 0.120) = 1/365.$$

Suppose also that levels from day to day are independent. Short-term (hour-to-hour), temporal correlation is to be expected, as is 'patterning' throughout the day. However, day-to-day relationships should be less in evidence, and any minor departures from independence should not seriously distort the following calculation.

Under these plausible assumptions we can calculate the probability of violation, that is, of having more than three occurrences of more than 0.120 ppm, when the true expected number of days is *just* over one, that is to say, the standard is formally being breached. This probability is 0.353. (Technical details are that over any period of three years the number of exceedances N should essentially follow a binomial distribution, Bin(3 × 365, 1/365), that is, be effectively Poisson with mean 3. Then $P(N > 3) = 0.353$.) This is an interesting result: if the standard is (just) breached there is only about a one in three chance of detecting the fact, using the specified criterion.

Consider some implications of this result. We might at first sight think that a one in three chance is unreasonably low − two-thirds of violations will not be picked up. Should we not ask for a compliance procedure with a much higher probability of detecting violation: say, 0.9 or 0.95? The problem is that even in its present form there is also a chance of up to one in three that a conforming agency *just* meeting the standard will be declared in 'in violation'. If the probability of detecting genuine violation were increased to 0.9 or 0.95, we would correspondingly increase the chance of incorrectly concluding that the standard is not met.

Thus the prescribed level of verification is not easy to declare. It may be that in some circumstances it should not be too high; to balance prospects of valid and of invalid declarations of violation, it might be 'even-handed' to choose a level of about 0.5. There is a major problem of balance here, not dissimilar to that encountered in dealing with 'producer's risk' and 'consumer's risk' in quality assurance, and an approach sometimes used in this context may be appropriate. If we are aiming at a level of 0.12 ppm, perhaps we should formulate the verification procedure to ensure on the one hand that if the level is 0.14 ppm (say) there is a high probability (0.95 perhaps) of detecting this, while on the other hand if the level is 0.10 ppm (say) this probability should be low (0.05 perhaps). Thus we protect both the environment by picking up marked violation and the conforming agency by maintaining a high probability (0.95) of avoiding false declaration of violation. Of course, the levels need to be carefully chosen as part of the specification of the verification procedure − in the ozone example, it may be that critical levels of 0.08 ppm and 0.12 ppm might be better (in order better to safeguard environmental protection against ozone damage). Thus specification of the level of the ideal standard, the form in which it is expressed − possibly as a pair of values (0.10, 0.14) rather than a single value 0.12 − and the nature of verification procedure all need to be balanced.

But let us return to the existing ozone standard. We should also ask how the level of 0.120 ppm was arrived at. One must assume that it was based on some analysis of the pollutant–effect relationship, but we are not aware of the aims of the standard being expressed in terms of what (changes in) effects on the human population are intended to be achieved by the standard. Nor are we aware of any statistical analysis of the extent to which the standard of not expecting to exceed an hourly maximum of 0.120 ppm on more than one day per year would contribute to such an aim.

In short, referring to the three aspects of uncertainty and variation presented in section 5.1, this example implies on aspect (a) an incomplete (at best) analysis of the pollutant–effect relationship to select the level of the standard, a sound attempt to deal with aspect (b) in regard to handling uncertainty and variation and, with respect to aspect (c), a clear verification procedure but one which is highly prescriptive and whose statistical properties were not explored (and which on investigation here seem to raise some serious problems). There is also confusion over the notion of 'expectation' or 'expected number'. In the standard it is clearly (and correctly interpretable as) the mean of a distribution. In the compliance criterion it is used as a sample average; indeed, Appendix H formally declares that 'the statistical term "expected number" is basically an arithmetic average'. Thus, in summary, we have statistically aware, statistically formulated, standard and compliance procedures but they leave some doubts about appropriate probabilistic and statistical measures and about the matching of the level of the standard and the level of assurance provided about meeting the standard.

5.2.3   LAND-BASED DISPOSAL OF LOW- AND INTERMEDIATE-LEVEL NUCLEAR WASTE

We reported in section 1.5 on the basic aim of this standard – that the risk to an individual in the long term (after closure and sealing of the disposal site) of death due to escape of radiation from the site should not exceed 10^{-6} per annum. We now consider this standard in some detail. The standard is first defined in DOE *et al.* (1984), and has been refined in a series of guidance documents since, the latest being Her Majesty's Inspectorate of Pollution (HMIP) *et al.* (1995) and the recent White Paper (DOE, 1995). In broad terms, the requirement for any prospective operator of such a facility is to use 'best practical means' (BPM) to reduce the risk to 'a representative member' of a homogeneous most exposed group of the general human population to below 10^{-6} per annum, or to a level 'as low as reasonably practicable' (ALARP) without incurring 'disproportionate costs'.

This is a standard set directly on the effect. At least superficially, it is far from being an exemplary case of uncertainty and variation being recognized fully in the setting of a standard. Consider the three criteria in the previous section.

(a) Since this is a standard on the effect, the benefit is immediate and obvious. However, those involved in setting the standard would have had no idea what the cost of complying with it would be. There was no analysis of the chain from actions through pollutant levels to effect. The value of 10^{-6} per annum has been used in other standards as a level of acceptable risk to human populations, and was not derived specifically for use in this standard. The cost side of the cost–benefit relation is dealt with, as in various other standards, by reference to the abbreviations BPM and ALARP, together with the phrase 'without incurring disproportionate costs', and these are clearly close relatives of the BATNEEC and ALARA terms discussed in section 2.9. Our criticism of such an approach is that it is vague and ducks the real question of balancing cost and benefit.

(b) The statement of the standard makes some recognition of inherent variation and uncertainty. It applies to the most exposed group of people rather than the general population, recognizing that there will be substantial variation, for instance between those living close to the disposal site and those living very far away. It recognizes that even within a 'homogeneous' group of people there will be variation, and the intention is not to protect the most extreme individuals, so the standard refers to 'a representative member' of such a group. However, this again ducks the serious questions of understanding and quantifying the uncertainty and variation, and leaves great potential ambiguity since terms such as 'homogeneous', 'group', 'most exposed' and 'representative' are not defined. For instance, is a 'representative' supposed to refer to the average risk of the group, its median, or what?

(c) The standard gives no guidance on how it might be verified in practice. There is no requirement at all on the quality, in statistical terms, of verification.

On the face of it, then, this fails on all counts to give adequate recognition of uncertainty and variation in standard-setting, but this would be a harsh judgement. The issues which have been evaded in headings (a) and (b) above are admittedly difficult, and the regulatory agencies have attempted to produce workable guidance. The fact that this is rather vague and ill-defined throughout is perhaps more acceptable in this case because of the considerations set out below, which basically rely on the fact that there are expected to be very few actual disposal sites proposed, and each will involve a very substantial investment. To date, only one such site has been approved in the UK, at Drigg in Cumbria, where British Nuclear Fuels Ltd (BNFL) has established a 'shallow' facility for low-level waste disposal. Proposals for a much more ambitious deep disposal site for low- and intermediate-level waste have been under development by Nirex for some years.

What makes this example particularly interesting and instructive is the huge effort which has been expended on procedures to verify that any proposed or operating disposal facility actually satisfies the standard. Much of this work has been carried out in the UK over a period of years, first by HMIP and more recently by the EA. The EA is charged with assessing any proposal for such facilities, and has developed extensive methodology in the process of assessing the BNFL Drigg site and in preparation for assessing the Nirex proposals. See Thompson and Sagar (1993) and other papers in Bonano and Thompson (1993), and more recently Thompson et al. (1996), for much valuable review material. This (and supplementary regulatory guidance over the years) has done much to answer our criticisms of the standard.

Primarily, this work comprises highly sophisticated modelling of the chain (see Figure 2.1) from design and implementation of the disposal facility (the 'actions'), through eventual escape of radionuclides into the ecosphere around the facility (the 'pollutant at entry locations') and the transport of those radionuclides to the ground surface, rivers, atmosphere (the 'pollutant at contact locations'), to radiation doses received by people living in the vicinity (the 'effects'). The complex of models is built on the best available scientific evidence. The models contain many uncertain physical parameters, and the EA has been active in the development of reliable procedures for elicitation of expert probability distributions for parameters, as have similar agencies in the USA and elsewhere (see, for example von Holstein and Matheson, 1978; O'Hagan, 1995).

The models are implemented in an extremely large and complex computer system, the core of which is a computer code known as VANDAL (Kane, 1992). Essentially, VANDAL carries out probabilistic risk analysis (or 'probabilistic system analysis', as EA now refers to it, or uncertainty analysis) using Monte Carlo technology. Using the various model components, together with probability distributions representing uncertainty about parameters within those models, it computes a probability distribution for the effect (long-term dosage to individuals). Since long time-scales are involved, simulation of the effects of climate-driven environmental changes is incorporated through another Monte Carlo model, called TIME4, which provides corresponding, time dependent boundary values for the models in VANDAL.

As noted in section 3.2.6, this probabilistic risk assessment is incomplete if it takes no account of uncertainty about the models upon which it builds. The EA approach to this kind of uncertainty is comparably sophisticated. An 'uncertainty and bias audit' is carried out on all components of the modelling to identify all the assumptions and approximations made at every stage. Alternative, plausible assumptions or approximations are identified where appropriate in this audit. There is then careful assessment of whether, and if so how, the conclusions of the analysis might be modified if alternative assumptions or approximations were used. Where sensitivity is thought to be potentially important, separate analyses might be conducted under those alternative model specifications. This falls short of a thorough analysis of model uncertainty, but is perhaps the best attempt so far at this difficult topic.

The effect of adding the substantial body of work on assessing compliance with this standard to the definition of the standard itself, is to produce one of the better examples that we know of in terms of recognition of uncertainty in setting environmental standards. There remain, however, a number of areas of concern and ambiguity.

(a) We see in this example a highly complex analysis used to ascertain compliance, but this might serve as an example of good practice generally in understanding the pollutant–effect relationship and probabilistic risk analysis. Such analysis should be employed in the determination of a standard in the first place, as an aid to understanding either or both of the benefit and cost of any proposed standard, rather than simply as a tool for assessing compliance. In the context of a very few, highly complex and individual, 'locations' for application of the standard, it is understandable that such an approach was not seen as necessary in the formulation of this particular standard, but we believe that this kind of sophisticated analysis should take place in support of determining the level at which to set standards generally. While it might not be practical to devote so much effort in most other problems of environmental pollution standard-setting, it shows that an important area for future research is to distil the essence of such an approach into usable guidelines for more routine applications.

(b) There is clearly scope for more thorough treatment of uncertainty in models. We might identify in this example an important step, which is that all the modelling is aimed at assessing the effective radioactive dose that would be received by a representative individual, and this is converted to a risk of fatal cancer by multiplication by a simple factor. Uncertainty in that factor is not at present accommodated in the modelling.

(c) One source of uncertainty about that factor is ambiguity over the meaning of the term 'risk' as used in the standard. We have pointed out in section 1.5.3 the diverse ways in which 'risk' is used, and a major deficiency in this standard is that it does not explain how the term is to be interpreted. It is perhaps intended that a 'risk of 10^{-6}' means in effect a probability of 10^{-6}, and this is how it seems to be interpreted in practice, but this should be clarified.

(d) The use of imprecisely defined critical groups and the concept of a representative member of a group are unsatisfactory. Guidance that has evolved in this problem has helped to clarify a 'most exposed' group in the near future after closure of the facility, but there is still considerable ambiguity over the definition of such groups in the far future. The analysis is supposed to examine risk up to 1 million years after closure (although this figure itself is no more than a generally accepted interpretation of a time-scale that is not laid down in the original standard), and identifying a 'most exposed' group so far into the future is fraught with difficulty. The difficulty is partly a result of inadequately allowing for random variation in the population.

(e) Finally, the quality of statistical verification of the standard should be specified. At present, the outcome of the probabilistic risk analysis is a probability distribution for the true risk of 'a representative member of a most exposed group' at any time in the future, representing uncertainty in the true risk arising from uncertainty over model parameters, etc. How high should the probability of a risk below 10^{-6} per annum be in order to say that the disposal facility is satisfactory (without having to argue that the risk is ALARP)?

So, despite the effort involved, but perhaps because of the complexity of this context, there remain a number of problems and ambiguities to resolve. The understanding of the many, often subtle, ways in which uncertainty and variation arise is a task for skilled statisticians, working closely with other scientists and the regulatory agencies.

5.3   EXAMPLES WHICH PAY NO REGARD FOR UNCERTAINTY AND VARIATION

There are many examples of standards that pay little or no regard whatsoever to the facets of uncertainty or variation which will be experienced in the practical situation where the pollutant will be found. Furthermore, many standards prescribe no statistical methodology or probabilistic level of assurance which must be achieved.

In a sense, realizable standards fall in the second category, since they consist merely of a statement of a sampling procedure that must be followed and a specific set of outcomes that must be achieved. Note, however, that these prescriptions could have been arrived at in the first place by detailed consideration of probabilistic and statistical analyses of the prevailing uncertainty and variation, or could subsequently be subjected to such an analysis to determine precisely what they imply about the situation 'at large'. This is exemplified by the sampling procedure used to test compliance with water-quality standards in respect of faecal coliforms proposed by ANZEC and described in section 4.3.1 above. Here 6 monthly samples have to be chosen and the median value must be below 150 c.f.u. and the two largest values must be below 160 c.f.u. This may be an appropriate procedure. But what does this realizable standard imply about the quality of the water in an overall sense? What assurances does it give? Was any ideal standard considered in arriving at such a sampling procedure? Such considerations could have been used to produce the more acceptable form of standard: namely a statistically verifiable ideal standard.

What we have termed an ideal standard is almost bound to fall in the first category. It consists of a set of circumstances that must precisely prevail, often expressed in the form of a maximum allowable level of a pollutant in a defined medium. For example, in DOE (1996), we read that 'environmental quality standards . . . concern the levels of pollution not to be exceeded in a given geographic area or medium, however many sources of emissions there may be in that area'. How can we ever know precisely if such a standard is being met? DOE (1996) contrasts such standards with alternative forms such as 'dose limits', 'process standards', 'emission or discharge standards', 'product standards' and 'management standards'. While some of these forms encounter lower degrees of intrinsic variability ('process', 'product' or 'management' versions), they are not free from such influences and, if expressed in ideal form, with no in-built procedure for statistical verification, remain equally sterile. This problem is particularly acute, for example, for some air-quality standards where 'certain fixed limits must be applied', and for pollution of agricultural land (for example, the Nitrates Directive limits the application of organic manure to 170 kg/h on the justification of reducing and preventing corresponding water pollution, but with no apparent detailed consideration of the indeterministic links and with recommendation for 'monitoring' only at a very coarse level). As another example, the Paris Commission limits values of SO_2, NO_2, chemical oxygen demand and suspended solids (for the sulphite paper pulp industry) with no indication of how such levels are to be confirmed in practice.

Another interesting example is that of the Expert Panel on Air Quality Standards (1995) recommendation on particulates. The Expert Panel examines physicochemical composition, sources of airborne particulates, measurement, monitoring, health aspects and the need for a standard which it concludes should at present take the form of limiting airborne particulates PM_{10} in the UK to a maximum of '50 µg/m³ measured as a 24-hour running average'. Just what is the nature of such a standard? Is it an ideal standard (to apply everywhere), or a realizable standard (to apply at a specific location)? How was the level determined? Was any direct account taken of uncertainty and variation or pollutant–effect relationships or costs and benefits in arriving at the level?

The ambiguous nature of ideal standards is clearly beginning to be recognized, since standards are more and more being augmented with an associated monitoring policy, such as, 'take and test samples to make sure that water meets the standards' or 'weekly samples of bathing water are taken' or 'a combination of point-source sampling and environmental monitoring . . . [is used] to ensure that consented parameters are not exceeded'. Where does such imprecise specification leave us, however: with a non-implementable standard and an inoperable vague principle for assessing compliance?

5.4 SUMMARY

We have attempted in this chapter to illustrate the different levels of incorporation of uncertainty and variation in the setting of pollution standards at the present time. In section 5.1 we have enunciated three criteria by which appropriate use of uncertainty and variation in standard-setting may be judged. Section 5.2 describes three current situations where standards make sound efforts to meet at least some of these criteria. In section 5.3, we illustrate how realizable and ideal standards can effectively pay almost no heed to such considerations.

Conclusions: the current situation and a look forward

6

In this report we have covered a wide range of topics related to the incorporation of uncertainty and variation in the setting of standards for environmental pollution. In conclusion, we need to draw together the strands by

- reviewing the broad principles for setting standards (section 6.1);
- summarizing the present situation in relation to such principles (section 6.2);
- proposing how the situation in the future should evolve towards setting standards which take proper account of uncertainty and variation (section 6.3); and
- providing a final overview of the key conclusions and recommendations of this report (Section 6.4).

6.1 BROAD PRINCIPLES FOR SETTING SOUND STANDARDS

Environmental pollution standards are intended to control environmental damage caused by polluting agents. We cannot hope to eliminate all harmful effects on individuals, on flora and fauna, on society, on ecosystems, etc., for a variety of reasons. Depending on circumstances, we may not fully understand what effects will occur and cannot assess or quantify benefits that would derive from control procedures. In any case, we could seldom afford the costs of required actions even if relationships between pollution levels, their effects and their benefits were known in any situation. Irrespective of such uncertainties, it is often unclear at what location it is best to set a standard for most efficient and cost-effective application: whether this should be in terms of actions at source (for example, by intervening in processes which potentially cause pollution), at a point where the pollutant enters the environment (controlling emission levels of specified pollutants in plant discharges) or at a point of contact with vulnerable targets (limiting nitrate levels in rivers).

In standard-setting there is scarcely anything that we can be absolutely certain of. We cannot be certain whether a standard needs to be set. If we choose to set a standard in a particular position, in a particular form with a particular level, we cannot know its consequences with certainty, in terms of either the actions and costs that will be induced in attempting to comply with the standard, or the effects and benefits for the environment and the subject group. Almost the only thing that is certain is that later, with the benefit of hindsight, we will see that the standard ought to have been set in a different way, at either a higher or lower level! Yet this also only serves to emphasize the uncertainties in standard-setting. Decisions must be taken, and standards set, in the face of uncertainty. If those actions are taken in full and careful recognition of the available knowledge (and lack of knowledge), uncertainty and variation, then they will represent the best decisions that could be made at that time. We cannot criticize them merely because at a later time, when we have more knowledge and less uncertainty, we see that better decisions could have been taken. The only fair criticism of a standard is if it has been set without detailed consideration of the consequences, including full and careful acknowledgement of uncertainty and variation.

A standard serves little purpose if we cannot assess its costs, effects and benefits and, in particular, if we cannot monitor practical compliance with the standard. At any stage in the process of attempts at assessment of cost, effects and benefits, and of associated choice of point of application, we encounter a burgeoning array of more fundamental forms of uncertainty and variation (variability) which conspire to make such assessment and choice difficult to effect, and must inevitably be taken into account if sound evaluation and choice are to be possible. These relate to the indeterminacy that is bound to be present. A pollutant will not have a unique determined effect. A whole range of types of effect and levels of effect can arise from one occasion to another (temporal variation), from one place to another (spatial variation), because of differential responses of a subject (intrinsic or natural variation) or imprecision in our means of measuring the effects (measurement error). The same considerations affect the pollution levels, as well as their effects; and the costs of implementing a pollution policy or the benefits it produces.

Uncertainties also beset our choice of what measures of pollutant and what forms of effect we are best to consider. Often there will be many relevant measures (or pollutants) and many relevant effects; and relationships will need to be considered between multivariate sets of measures and effects.

To handle the array of uncertainties and variational elements, we need to make use of the range of probabilistic and statistical concepts and methods, applying them at all stages in the study of relationships between costs, pollutant levels, pollutant effects and resulting benefits of any environmental control policy.

We have covered all these matters in our progression towards a proper basis for evaluation of the appropriateness and properties of current standard-setting procedures and in attempting to offer a sound policy for standard-setting in the future. Let us briefly review this process as presented throughout the report.

The introductory Chapter 1 begins with definitive matters, explaining the basic nature of such terms as pollutant, medium, location, sample, sample statistics, subject group and effect, without which we could not start to examine the complexities of the problem. It continues by drawing distinctions between different kinds of standard, applied to the level of effect on a subject group, to pollutant level in a medium or to actions which generate pollution, and it makes a crucial distinction (developed later) between whether we express the standard in terms of absolute measures in the environment or observed values in a sample. The pollutant–effect relation is (as explained above) central to standard-setting, and this is employed as a basis for delimiting the various forms of uncertainty and variation encountered. The closing section of the chapter covers the vital question of how we represent uncertainty and variation in operational terms, explaining basic ideas and interpretations of probability and other concepts relevant to this matter.

Chapter 2 is central to the theme of this review section. It covers the range of topics which have to be considered in any standard-setting approach sensitive to variation and uncertainty. There is a natural progression from the costs which underlie any pollution-control process to the benefits that ultimately flow from it. In attempting to relate standards to initial broad aims, refined into specific objectives, it is important to develop the notion of progression in terms of what we have termed the cost–benefit chain, in which we observe the important roles played by the stages at which standards might be set – namely, in respect of basic actions on the process generating the pollution; in respect of outcomes (levels) at initial entry of the pollutant into the system; in respect of outcomes (levels) affecting the vulnerable agents in the system; and in terms of overall effects on the system. In all cases an understanding of the pollutant–effect relationship (expressed in regard to uncertainties and variability) is vital, as is a recognition of the varying degrees of uncertainty that also arise from standards being set at various levels of separation from costs at the one extreme, or from benefits at the other.

This leads to a fundamental distinction between ideal standards which describe the 'global' features of the system and embody no formal means for monitoring compliance, and realizable standards expressed in terms of observations of sample behaviour in practice but which (although intrinsically self-monitoring) in their basic form provide no formal way of inferring their global effect. The introduction of sampling is essential, but adds a further dimension of influence of uncertainty and variation.

At this level of broad principles, it is relevant to contrast flexibility and complexity. Flexibly expressed standards with multiple objectives are attractive but are bound to make standards more complex in form. Simply expressed standards limit the flexibility of their application and the extent to which they reflect overall influence on the system. Ideal standards can be readily set in flexible form but are difficult to monitor. Realizable standards can be simple in form but provide limited information about the overall, global situation. Neither form is acceptable in general as a basis for setting standards.

To harness uncertainty and variation in the standard-setting problem requires methods for interpreting and analysing information subject to them. Thus we need to understand statistical approaches and procedures which utilize probabilistic descriptions of a problem and enable measured or quantified inferences to be drawn about the underlying (overall or global) situation. We have explained basic concepts of statistical inference and the central role of statistics in the standard-setting process. This is related back to the uncertainty and variation encountered at various stages of the cost–benefit chain. Simple examples are discussed throughout to illustrate the rather subtle concepts and processes that arise.

This brings us naturally to the crux of the problem. In what form should standards be set? Realizable standards are superficially attractive in that we can immediately exhibit compliance, or violation. But they must be extended to the drawing of appropriate statistical inferences about the underlying situation – to quantified statements with associated measures of probabilistic assurance about that situation, such as 'if we just satisfy the standard then with probability 0.95 the actual amount of the pollutant in the environment will be no more that some level x, or the incidence of some effect will be no more than some level y.' But these levels x and y may not be what we want, or would have required in the first place.

Ideal standards allow us to delimit such global levels, but give us no means of assessing compliance. They in turn need to be augmented into what we term statistically verifiable ideal standards, which comprise a joint prescription of a level 'at large' (in the global situation) which should not be exceeded together with a requirement to demonstrate statistically a prescribed degree of assurance – a probability of (say) 0.9 that the true level specified is not being exceeded (or one of many other statistical means of expressing such assurance).

This is the form of standard which we recommend. It addresses the global situation and embodies its own statistical/probabilistic compliance criteria. Note the vital non-prescriptive element in this approach. Any sound means of demonstrating the required level of assurance should be acceptable. No specific sampling approach or statistical methodology is mandated; merely the level of assurance of any adopted compliance policy. This allows a vital degree of flexibility for the 'complier' to act reasonably (cost-effectively) in seeking to satisfy the standard and in demonstrating that this is being done.

The question of multiple standards is addressed. How do we compare standards set in different terms, at different stages of the cost–benefit chain, for the same set of objectives? We pose this as an important area for flexibility and further study.

If we approach standard-setting (as presumably we should aim to do) on the basis of fundamental cost–benefit balance, rather than in intermediate terms of actions, or levels of pollutant or its effect, many other factors need to be incorporated, quantified and analysed. This is necessary for the issue of determining what level should be specified in the ideal standard. The methods of decision theory provide a means of doing this, incorporating formal notions of utility, probability and risk (different from what is often rather superficially described as 'risk analysis'). We outline this approach but note that its complexity and cost (and added uncertainties) will often be such that it is hard to justify such an extended methodology for setting standards.

In view of the fact that knowledge of the pollutant–effect relationship is central to any reasonable approach to setting standards, we develop this theme extensively in Chapter 3. It is important to recognize, however, that pollutant–effect relationships are studied here as representative of various links in the chain from actions to effects. In order to assess the proper position, form and level of a standard, we need to understand the relationships between actions and pollutant levels where the pollutant enters the medium, between pollutant levels at these entry locations and at locations where the medium comes into contact with the vulnerable subject group, as well as between these pollutant levels and the effects on the subject group. To study such relationships requires an understanding of the importance and nature of (statistical, but also mechanistic) models to express the relationship and of appropriate (statistical) methods to analyse it. This is explained in some detail. To add important substance to the general situation, we review the literature of the last few years on specific examples of many different practical problem areas, and of the wide range of statistical models and methods used to study them. In covering major topics such as sampling methods, spatial temporal models, dose-response procedures, the Bayesian approach, nonparametric and multivariate methods, assessment of extreme values and use of deterministic (mechanistic) models, we seek to provide a detailed overview of approaches to examining the pollutant–effect relationship for detailed application to pollution standard-setting.

6.2   THE PRESENT SITUATION

A natural question to ask is the following. If the above principles are those which should underlie the setting of standards for environmental pollution, is this what we find when we examine the present situation? Unfortunately, the answer is clearly *no!* We encounter:

- relatively little acknowledgment, let alone incorporation, of uncertainty and variation;
- few (if any) attempts to construct standards in terms of well-argued general principles related to aim, objective, cost and benefit, or point of entry of the pollutant to the system;
- a preoccupation with ideal standards, frequently restricted to pollutant levels, and little concern for implementation or compliance procedures;
- a proliferation of informally (vaguely) expressed general principles such as 'safe levels', 'the precautionary principle' ALARA, BATNEEC, 'extrapolation', critical levels and loads, critical groups, etc., which while unobjectionable in intent do not really provide a basis for reasoned setting of explicit sound standards.

We have covered these points as appropriate at various stages of the report and have exemplified (in Chapter 5) the different levels of sophistication, from the rare case of clear recognition and incorporation of uncertainty and variation to the more common situation where such essential ingredients are ignored.

It is useful at this stage to review where current policies and approaches are described in the report, and to draw attention to emerging (more hopeful) trends.

In section 2.9 we have examined the range of intuitively appealing broad principles which currently underlie the setting of standards. These include attempts to achieve 'safe' levels, or 'prudent reduction' from the prevailing situation; and the use of so-called 'precautionary principles'. These embody no overt consideration and employment of costs, benefits, uncertainty or variability. More motivated by (but again not designed formally in consideration of) costs are the principles of BATNEEC, BPEO and ALARA. While the basic stimulus of such approaches has some desirable characteristics, they hardly constitute structured procedures for setting standards in the face of the complex aims and objectives and the ubiquitous presence of uncertainty and variation.

More structured in form are the emphases outlined in section 3.5, which include the (rather ill-defined) notion of environmental impact assessment, and the use of critical groups or of critical levels and critical loads. Again, however, they are not expressed formally enough to lead to the required detailed analyses of cost, benefits, effects, uncertainty and variation that should support any sound approach to standard-setting. The difficulties of combining information and of extrapolation are also discussed.

In Chapter 4 we review more encouraging prospects, in terms of work that is presently being conducted on detailed statistical studies of pollution effects in the contexts of standards, and of the way in which attitudes are changing in recognition of the need properly to incorporate uncertainty and variation.

6.3 A LOOK FORWARD

We have seen growing sophistication of the processes of setting standards, and a growing awareness of the importance of uncertainty and variation. The commissioning of this report is clear evidence of the latter trend, which surely will and should continue. We have tried herein to identify appropriate ways to account for uncertainty and variation in setting environmental pollution standards. It is our hope that the continuation of current trends will result in future standard-setting increasingly conforming to the guidelines outlined below and set out in detail in this report. However, it will surely be necessary to encourage this by promoting the conclusions and recommendations of the report, and by seeking to set up research and development studies which are designed to produce standards in specific cases exemplifying the good practices which we have proposed.

6.3.1   FORMAT OF STANDARDS

We have argued in favour of statistically verifiable ideal standards. These combine a statistically based ideal standard for levels of pollutant or for effects (recognizing natural variation and uncertainty) with a standard for the quality of statistical verification of the ideal standard. Although we are not aware of any instance of such a standard at present, we believe that it will be found to be the most appropriate form of standard in future. We stress the importance of uncertainty and variation, which are explicitly recognized in two ways in such a standard. The importance of recognizing variation in the statement of the ideal standard is increasingly understood, and we have identified several good examples. However, the ideal standard is typically accompanied at present either by no specification at all of the procedures for statistically verifying the standard (on the basis of sampling), or by prescribing a particular sampling scheme (often without statistical analysis of how well the sampling process really succeeds in verifying the ideal standard).

We believe that the statistically verifiable ideal standard is the best form, because it avoids being over-prescriptive, while not making the mistake of leaving the standard open to statistically inadequate (or no) verification.

6.3.2   SETTING THE LEVEL

In addition to the form of a standard, we must specify the 'content', the level of the standard which determines how stringent it is. For a statistically verifiable ideal standard we must specify the required pollutant or effect level (and whether this is an average, 95% point or whatever). A more stringent standard will in general achieve greater benefits, but entails greater costs. Setting the level therefore inevitably involves striking a balance between cost and benefit. We believe that this balance should be struck by an open, explicit assessment of both costs and benefits and preferably by means of an objective decision-theoretic cost–benefit analysis.

The statistical verification criterion of a statistically verifiable ideal standard must be set as an intrinsic part of setting the level of the ideal standard. It requires attention to the strength or quality of statistical verification to be demanded, and the extent to which the benefit of the doubt (or onus of proof) is given to the polluter or to the environmental or regulatory agency. (The discussion of the US ozone standard in section 5.2.2 shows how this will typically be linked to the level of the ideal standard.) A less strict statistical verification standard weakens the statistically verifiable ideal standard as a whole, for example, and may need to be compensated by a tightening of the level of the ideal standard.

In order to assess cost and/or benefit, no matter what form of standard is chosen and where it is placed, it is necessary to understand the pollutant–effect relationship (and the action–pollutant relationship, for which identical considerations apply). Such relationships may be represented using theoretical, mechanistic models or empirical, statistical models, and are typically best presented through a (possibly complex) hybrid of both these kinds of component. However the modelling is structured, uncertainty and variation will be pervasive. It is essential to give proper recognition to uncertainty about both the benefits arising from setting a standard at any particular level and the costs of achieving compliance with the standard.

The great variety of modelling approaches used in different contexts renders it difficult to make specific recommendations about the proper treatment of uncertainty and variation in modelling the action–pollutant—effect relationships to assess costs and benefit. The following are a few general points.

- Modelling which ignores sources of uncertainty and variation is unacceptable, because it gives a false (and potentially dangerous) impression of precision.
- An enormous variety of statistical models are found to be appropriate for environmental problems. Many are sophisticated, and the corresponding statistical analysis demands careful handling by a qualified statistician. Inappropriate statistical methods can lead to seriously erroneous conclusions.
- In order to accommodate and combine all the various components of uncertainty and variation, probabilistic risk analysis or detailed uncertainty analysis are the appropriate tools (in the respects explained in this report).

6.3.3   THE NEED FOR STATISTICIANS

We believe that the above proposals genuinely represent the way forward in setting environmental pollution standards. Accordingly, there will be a need for sources of substantial support for those working in environmental problems, with specialist knowledge of statistical techniques appropriate to the setting of environmental pollution standards. While environmental statistics is a growing branch of statistical study (as witnessed by the Royal Statistical Society's recent creation of an Environmental Statistics Study Group), the depth of expertise available at present will nowhere near satisfy that need. When high-quality statistical advice is not readily available, it is all too easy to settle for the simplistic recipes of off-the-shelf statistical software, and misunderstanding of the true uncertainties is an almost inevitable consequence. Action is needed both to impress the need for expert statistical advice on those responsible for setting environmental pollution standards, and to raise the profile of environmental statistics in order to encourage the provision of an adequate supply of expertise.

6.4   OVERVIEW

Finally, we seek here to encapsulate the major conclusions of our report.

1. Uncertainty and variation in many forms permeate all aspects of environmental pollution standard-setting problems. Standards must be set in ways which give full recognition to all sources and forms of uncertainty and variation.

2. The proper quantification of uncertainty and variation is through probability. The methods of statistics are essential for measuring probabilities and for making inferences and decisions in the face of uncertainty and variation. Probability and statistical analysis should be employed in standard-setting to recognize, quantify, analyse and understand the impact of uncertainty and variation.

3. Standards can be characterized in terms of position, form and level. The form of a standard must be set so as to take account of uncertainty and variation, otherwise the standard will be meaningless, unenforceable or indefensible. Apart from exceptional cases where a realizable standard may usefully be set on direct actions to control pollution, the only acceptable form of standard is a statistically verifiable ideal standard. The proposed form of standard is the statistically verifiable ideal standard. This combines an ideal standard with a standard for the quality of statistical verification of compliance with the ideal standard.

4. There appear to be no examples of statistically verifiable ideal standards in present practice. Many current standards fall so far short of proper recognition of uncertainty and variation that they are deeply flawed. Consideration should be given to identifying inadequate current standards, and to reformulating them as statistically verifiable ideal standards.

5. The stringency of a statistically verifiable ideal standard is determined by the ideal standard, through the choice of the feature(s) of the probability distribution to control and through the level at which a feature is controlled, and by the standard for statistical verification of the level. These can only be set by appropriate statistical analysis which balances the costs and benefits. The setting of the level and statistical verification requirements of a standard must include detailed recognition of costs and benefits, through appropriate statistical analysis. Although a full decision-theoretic cost–benefit analysis is rarely practicable, actual analysis should be guided by its principles.

6. Proper analysis of costs and benefits demands an understanding of the various links from actions to effects. Chief among these are relationships between pollutant and effect, but all these links require appropriate modelling and recognition of uncertainty and variation. Understanding of pollutant–effect and other links from actions to effects should be central to the setting of the level of a standard. Models must be validated wherever possible. The many sources of variation and uncertainty should be recognized and should be analysed by means of probability models (relational, temporal, spatial) and associated statistical inference procedures. It is essential to recognize all the sources of uncertainty in a model, for which tools such as uncertainty analysis are important.

7. Statistically verifiable ideal standards, the significance of the chain from actions to effects, and the ideas of consistency of standards at different positions on that chain, are all new constructs. There is scope for considerable research and development, particularly in regard to drawing together all the strands of this report in specific case studies. There is a need to promote projects of research

and development into statistically verifiable ideal standards, consistency of standards and the proper setting of levels. Paradigm cases to exemplify good practices are urgently required.

8. Proper recognition of uncertainty and variation can only be implemented throughout the many areas of environmental standard-setting if there is an adequate source of high-quality statistical expertise. The statistical community is increasingly coming to appreciate the importance of environmental statistics, but suitably qualified statisticians with experience in environmental problems and interdisciplinary skills will only arise if their importance is recognized by the regulatory agencies. Statistical expertise should be seen by environmental standard-setting and regulatory agencies to be as necessary as scientific expertise in other disciplines, such as chemistry, engineering or physics. Just as the agencies demand well-qualified specialists in those disciplines, they should be seeking to recruit from the most able statisticians.

9. The above proposals are far-reaching. Although we detect encouraging signs that practices are already moving in the right direction, it would be easy to underestimate the degree of culture change required. The need for proper statistical handling of uncertainty and variation must be truly acknowledged by those involved in setting environmental standards.

References

Abraham, H.J., Schlums, C. and Frenzel, W. (1993). Interlaboratory test for external quality control. 4. Statistical evaluations and assessments of results. *Staub-Reinhaltung der Luft*, **53**, 355–361.

Adams, S.M., Ham, K.D. and Beauchamp, J.J. (1994). Application of canonical variate analysis in the evaluation and presentation of multivariate biological response data. *Environmental Toxicology and Chemistry*, **13**, 1673–1683.

Ahlfeld, D.P. and Islam, M.S. (1994). Estimating the probability of exceeding groundwater quality standards. *Water Resources Bulletin*, **30**, 623–629.

Aldenberg, T. and Slob, W. (1993). Confidence limits for hazardous concentrations based on logistically distributed NOEC toxicity data. *Ecotoxicology and Environmental Safety*, **25**, 48–63.

Altshuller, A.P. and Lefohn, A.S. (1996). Background ozone in the planetary boundary layer over the United States. *Journal of the Air and Waste Management Association*, **46**, 134–141.

Amin, M.B. and Husain, T. (1994). Kuwaiti oil fires – air-quality monitoring. *Atmospheric Environment*, **28**, 2261–2276.

Aruga, R., Negro, G. and Ostacoli, G. (1993). Multivariate data analysis applied to the investigation of river pollution. *Fresenius Journal of Analytical Chemistry*, **346**, 968–975.

Ashenden, T.W., Bell, S.A. and Rafarel, C.R. (1996). Interactive effects of gaseous air-pollutants and acid mist on 2 major pasture grasses. *Agriculture Ecosystems and Environment*, **57**, 1–8.

Ashmore, M.R., Bell, J.N.B. and Brown, I.J. (eds) (1990). *Air Pollution and Forest Ecosystems in the European Community*, EU Air Pollution Research Report No 29. Brussels.

Attrill, M.J., Rundle, S.D. and Thomas, R.M. (1996). The influence of drought-induced low fresh-water flow in an upper-estuarine macroinvertebrate community. *Water Research*, **30**, 261–268.

Barnett, V. (1976). The statistician: Jack of all trades, master of one? *The Statistician*, **XXV**, 261–278.

Barnett, V. (1982). *Comparative Statistical Inference*, 2nd edn. Wiley, Chichester.

Barnett, V. (1997). Statistical analyses of pollution problems. In Barnett, V. and Turkman,

K.F. (eds) *Statistics for the Environment 3: Pollution Assessment and Control*. Wiley, Chichester.

Barnett, V. and Lewis, T. (1967). A study of low-temperature probabilities in the context of an industrial problem (with Discussion). *Journal of the Royal Statistical Society A*, **130**, 177–206.

Barnett, V. and Moore, K.L. (1997). Best linear unbiased estimates in ranked-set sampling, with particular reference to imperfect ordering. *Journal of Applied Statistics*, **24**.

Barnett, V. and Turkman, K.F. (eds) (1993). *Statistics for the Environment*. Wiley, Chichester.

Barnett, V. and Turkman, K.F. (eds) (1994). *Statistics for the Environment 2: Water-related Issues*. Wiley, Chichester.

Barnett, V. and Turkman, K.F. (eds) (1997). *Statistics for the Environment 3: Pollution Assessment and Control*. Wiley, Chichester.

Barnett, V., Landau, S., Colls, J.J., Craigon, J., Mitchell, R.A.C. and Payne, R.W. (1997). Predicting wheat yields: the search for valid and precise models. To appear in *Precision Agriculture: Spatial and Temporal Variability of Environmental Qualities*, Ciba Foundation Symposium 210. Wiley, Chichester.

Becka, M., Bolf, H.M. and Urfer, W. (1993). Statistical evaluation of toxicokinetic data. *Environmetrics*, **4**, 311–322.

Biggeri, A. and Marchi, M. (1995). Case–control designs for the detection of spatial clusters of disease. *Environmetrics*, **6**, 385–393.

Bonano, E.J. and Thompson, B.G.J. (eds) (1993). Special Issue on Probabilistic Risk Assessment on Radioactive Waste. *Reliability Engineering and System Safety*, **42**(2–3).

Boswell, M.T., Gore, S.D., and Patil, G.P. (1996). Annotated bibliography of composite sampling. Part A: 1936–92. *Environmental and Ecological Statistics*, **3**, 1–50.

Brown, P.J., Le, N.D. and Zidek, J.V. (1994). Multivariate spatial interpolation and exposure to air pollutants. *Canadian Journal of Statistics*, **22**, 489–509.

Burnett, R. and Krewski, D. (1994). Air-pollution effects on hospital admission rates, and the random effect modelling approach. *Canadian Journal of Statistics*, **22**, 441–458.

Butlin, R.N., Yates, T.J.S., Murray, M. and Ashall, G. (1995). The United Kingdom national materials exposure program. *Water, Air and Soil Pollution*, **85**, 2655–2660.

Calman, Sir Kenneth C. (1996). Cancer: science and society and the communication of risk. *British Medical Journal*, **315**, 799–802.

Cape, J.N. (1993). Direct damage to vegetation caused by acid rain and polluted cloud – definition of critical levels for forest trees. *Environmental Pollution*, **82**, 167–180.

Carroll, R.J. and Ruppert, D. (1984). Power transformation when fitting theoretical models to data. *Journal of the American Statistical Association*, **79**, 321–328.

Carstensen, J., Madsen, H., Poulsen, N.K. and Nielsen, M.K. (1994). Grey box modelling in two time domains of a waste water pilot scale plant. *Environmetrics*, **4**, 187–208.

Corey, P.N., Lou, W.Y. and Brodes, I. (1994). Analysis of the relationship between air pollution and asthma. *Canadian Journal of Statistics*, **22**, 459–470.

Crump, K.S. (1984). A new method for determining allowable daily intakes. *Fundamental and Applied Toxicology*, **4**, 854–871.

Crump, K.S., Allen, M.S. and Faustman, E. (1995). *The Use of the Benchmark Dose Approach in Health Risk Assessment*. Risk Assessment Forum, EPA, Washington, DC.

Crump, K.S., Guess, H.A. and Deal, K.L. (1977). Confidence intervals and test of hypotheses concerning dose response relations inferred from animal carcinogenicity data. *Biometrics*, **33**, 437–451.

Czaplewski, R.L., Reich, R.M. and Bechtold, W.A. (1994). Spatial autocorrelation in growth of undisturbed natural pine stands across Georgia. *Forest Science*, **40**, 314–328.

Daniel, T.C., Edwards, D.R. and Sharpley, A.N. (1993). Effect of extractable soil surface phosphorus on runoff water quality. *Transactions of the ASAE*, **36**, 1079–1085.

Department of the Environment (1991). *IPC Practical Guide/The Environmental Protection Regulations 1991*. DoE, London.

Department of the Environment (1995). *Review of Radioactive Waste Management Policy – Final Conclusions*, Cm. 2919. HMSO, London.

Department of the Environment (1996). *Evidence to Royal Commission on Environmental Pollution*. DoE, London.

Department of the Environment, Scottish Office, Welsh Office, Department of the Environment for Northern Ireland and Ministry of Agriculture, Fisheries and Food (1984). *Disposal Facilities on Land for Low and Intermediate Level Radioactive Waste, Principles for Protection of the Human Environment*. HMSO, London.

Draper, D., Gaver Jr, D.P., Goel, P.K. (1992). *Combining Information*. National Academy, Washington, DC.

Edland, S.D. and van Belle, G. (1994). Decreased sampling costs and improved accuracy with composite sampling. In Cothern, C.R. and Ross, N.P. (eds) *Environmental Statistics, Assessment, and Forecasting*, Lewis, Boca Raton, FL.

Englund, E.J. and Heravi, N. (1993). Multi-phase sampling for soil remediation. *Journal of Environmental and Ecological Statistics*, **1**, 247–263.

Espegren, M.L., Pierce, G.A. and Halford, D.K. (1996). Comparison of risk for pre-remediation and post-remediation of uranium undertakings from vicinity properties in Montecello, Utah. *Health Physics*, **70**, 556–558.

European Union (1994). Proposal for a Council Directive concerning the quality of bathing water (94/C 112/03). *Official Journal of the European Communities*, No C 112/3–10.

Expert Panel on Air Quality Standards (1995). *Particles*. HMSO, London.

Ferguson, C. (1992). The statistical basis for spatial sampling of contaminated land. *Ground Engineering*, June.

Ferguson, C. and Denner, J. (1994). Developing guideline (trigger) values for contaminants in soil: underlying risk analysis and risk management concepts. *Land Contamination and Reclaim*, **2**, 117–123

Ferguson, C. and Denner, J. (1997). Risk assessment using UK guideline values for contaminants in soil. In Lerner, D.N. and Walton, N.W. (eds) *Contaminated Land and Groundwater – Future Directions*, Special Publication. Geological Society, London.

Ferguson, C.C., Krylov, V.V and McGrath, P.T. (1995). Contamination of indoor air by toxic soil vapours: a screening risk assessment model. *Building and the Environment*, **30**, 375–383.

Gallegos, D.P., Webb, E.K., Davies, P.A. and Conrad, S.H. (1996). A probabilistic environmental decision support framework for managing risk and resources. In

Cacciabue, P.C. and Papazoglou, I.A. (eds) *Probability Safety Assessment and Management '96*, Vol. 1. Springer-Verlag, London.

Gardner, M.J. (1993). Investigating childhood leukaemia rates around the Sellafield nuclear plant. *International Statistical Review*, **61**, 231–244.

German Bundestag (ed.) (1989). *Protecting the Earth's Atmosphere*. Bonn.

German Bundestag (ed.) (1990). *Protecting the Tropical Forests*. Bonn.

Gonzalez Manteiga, W., Prada Sanchez, J.M., Cao, R., Garcia Jurado, I., Febrero Bande, M. and Lucas Dominguez, T. (1993). Time-series analysis for ambient concentrations. *Atmospheric Environment A*, **27**, 153–158.

Graf-Jaccottet, M. (1993). A flexible model for ground ozone concentrations. *Environmetrics*, **4**, 22–37.

Grimalt, J.O., Canton, L. and Olive, J. (1993). Source input elucidation in polluted coastal systems by factor analysis of sedimentary hydrocarbon data. *Chemometrics and Intelligent Laboratory Systems*, **18**, 93–109.

Gupta, R.C. and Albanese, R.A. (1994). Survival analysis of radiated animals incorporating competing risks and covariates. *Environmentrics*, **5**, 365–379.

Guttorp, P., Le, N.D., Sampson, P.D. and Zidek, J.V. (1993). Using entropy in the redesign of an environmental monitoring network evaluation. In Patil, G.P., Rao, C.R. and Ross, N.P. (eds) *Multivariate Environmental Statistics*. North-Holland/Elsevier Science, New York, pp. 175–202.

Guttorp, P., Meiring, W. and Sampson, P.P. (1994). A space-time analysis of ground level ozone data. *Environmetrics*, **5**, 241–254.

Harrison, P. (1992). *The Third Revolution*. I.B. Tauris/Penguin, London.

Haygarth, P.M., Fowler, D., Sturup, S., Davison, B.M. and Jones, K.C. (1994). Determination of gaseous and particulate selenium over a rural grassland in the UK. *Atmospheric Environment*, **28**, 3655–3663.

Haylock, R.J. and O'Hagan, A. (1996). On inference for outputs of compulationally expensive algorithms with uncertainty on the inputs. In Bernardo, J.M., Berger, J., Dawid, A.P. and Smith, A.F.M. (eds) *Bayesian Statistics 5*. Oxford University Press, Oxford, pp. 629–637.

Health Council of the Netherlands: Committee on the Evaluation of the Carcinogenicity of Chemical Substances (1994). Risk assessment of carcinogenic chemicals in the Netherlands, *Regulatory Toxicology and Pharmacology*, **19**, 14–30.

Her Majesty's Inspectorate of Pollution, Scottish Office, Welsh Office, Department of the Environment for Northern Ireland and Ministry of Agriculture Fisheries and Food. (1995). *Radioactive Substances Act 1995: Consultative Document: Disposal Facilities on Land for Low and Intermediate Level Radioactive Waste: Guidance and Requirements for Authorisation* (revised issue October). HMSO, London.

Hettelingh, J.P., Posch, M., Desmet, P.A.M. and Downing, R.J. (1995). The use of critical loads in emission reduction agreements in Europe. *Water, Air and Soil Pollution*, **85**, 2381–2388.

Heuss, J.M. and Wolff, G.T. (1993). Measurement needs for developing and assessing ozone-control strategies. *Water, Air and Soil Pollution*, **67**, 79–92.

Hinchee, R.E. and Olfenbuttel, R.F. (eds) (1991a). *In Situ Bioreclamation Applications and Investigations for Hydrocarbon and Contaminated Site Remediation*. Butterworth-Heinemann, Boston.

Hinchee, R.E. and Olfenbuttel, R.F. (eds) (1991b). *On-Site Bioreclamation Processes for Xenobiotic and Hydrocarbon Treatment*. Butterworth-Heinemann, Boston.

Hipel, K., Yin, X. and Kilgour, D.M. (1995). Can a costly reporting system make environmental enforcement more efficient? *Stochastic Hydrology and Hydraulics*, **9**, 151–170.

Holdren, G.R., Strickland, T.C., Shaffer, P.W., Ryan, P.F., Ringold, P.L. and Turner, R.S. (1993). Model selection and regionalization schemes. *Journal of Environmental Quality*, **22**, 279–289.

Howson, C. and Urbach, P. (1996). *Scientific Reasoning: the Bayesian Approach*. Open Court, Chicago.

Johnson, G.D., Nussbaum, B.D., Patil, G.P. and Ross, N.P. (1995). Innovative statistical mind sets and novel observational approaches to meet the challenges in the management of hazardous waste sites. In Lewis, R.A. and Subklew, G. (eds) *Challenges and Innovations in the Management of Hazardous Waste*. Air and Waste Management Association, Pittsburgh.

Kadane, J. B. and Wolfson, L. J. (1996). Experiences in elicitation. Research report, Carnegie-Mellon University Statistics Department.

Kane, P. (1992). *VANDAL Version 1.3 Technical Overview*, DOE Report DOE/RR/92.095. DOE, London.

Kaur, A., Patil, G.P., Sinha, A.K., and Tallie, C. (1995). Ranked set sampling: an annotated bibliography. *Environmental and Ecological Statistics*, **2**, 25–54.

Kerr, D.R. and Meador, J.P. (1996). Modeling dose response using generalised linear models. *Environmental Toxicology and Chemistry*, **15**, 395–401.

Kiilerich, O. and Ruff, H. (1994). Contaminated sites evaluation by use of the successive principle and artificial neural networks. *IFIP Transactions B*, **16**, 157–167.

Korn, L.R., Murphy, E.A. and Zhang, Z. (1994). Combining Wilcoxon tests with censored data: an application to well water contamination. *Environmetrics*, **5**, 463–472.

Küchenhoff, H. and Thamerus, M. (1996). Extreme value analysis of Munich air pollution data. *Environmental and Ecological Statistics*, **3**, 127–141.

Lacey, R.F., Gunby, A. and Hay, S.V. (1995). Methods of assessing compliance with standards for the quality of bathing waters. WRc Report DOE 3837/1.

Landner, L. (1994). How do we know when we have done enough to protect the environment? *Marine Pollution Bulletin*, **29**, 593–599.

Lawson, A.B. (1993). On the analysis of mortality events associated with a prespecified fixed point. *Journal of the Royal Statistical Society A*, **156**, 363–377.

Lawson, A.B. and Williams, F.L.R. (1994). Armadale – a case study in environmental epidemiology. *Journal of the Royal Statistical Society A*, **157**, 285–298.

Lee, E.H., Hogsett, W.E. and Tingey, D.T. (1994). Attainment and effects issues regarding alternative secondary ozone air-quality standards. *Journal of Environmental Quality*, **23**, 1129–1140.

Liepmann, D. and Stephanopoulos, G. (1985). Development and global sensitivity analysis of a closed ecosystem model. *Ecological Modelling*, **30**, 13–47.

Little, M.P. (1995). Are two mutations sufficient to cause cancer? Some generalisations of the two-mutation model of carcinogenesis of Moolgavkar, Venzon and Krudson, and of the multi-stage model of Armitage and Doll. *Biometrics*, **51**, 1278–1291.

MacGarvin, M. (1995). The implications of the precautionary principle for biological monitoring. *Helgolander Meeresuntersuchungen*, **49**, 647–662.

MacKenzie, A.B. and Scott, R.D. (1993). Sellafield waste radionuclides in Irish Sea intertidal and salt-marsh sediments. *Environmental Geochemistry and Health*, **15**, 173–184.

Madruga, M.R., Perera, C., Ade, B. and Rabello-Gay, M.N. (1994). Bayesian dosimetry: radiation dose versus frequencies of cells with aberrations. *Environmetrics*, **5**, 47–56

Malm, O., Castro, M.B., Bastos, W.R. *et al.* (1995). An assessment of Hg pollution in different goldmining areas, Amazon Brazil. *Science of the Total Environment*, **175**, 127–140.

McLachlan, D.R.C. (1995). Aluminium and the risk of Alzheimer's disease. *Environmetrics*, **6**, 233–275.

McNeney, B. and Petkau, J. (1994). Overdispersed Poisson regression models for studies of air pollution and human health. *Canadian Journal of Statistics*, **22**, 421–440.

Mikhailov, A.A., Suloeva, M.N. and Vasileva, E.G. (1994). Atmospheric corrosivity in cities and industrial centers of the former Soviet Union with respect to carbon-steel, zinc and copper. *Protection of Metals*, **30**, 329–335.

Moolgavkar, S.H., Dewanji, A. and Venzon, D.J. (1988). A stochastic two stage model for cancer risk assessment. I. The hazard function and the probability of tumor. *Risk Analysis*, **8**, 383–392.

Moore, D.R.J. (1996). Perspective: using Monte Carlo analysis to quantify uncertainty in ecological risk assessment. Are we gilding the lily or bronzing the dandelion? *Human and Ecological Risk Assessment*, **2**(4).

Muirhead, C.R. (1996). Statistical practice in the setting of standards in radiological protection. Personal communication.

Muirhead, C.R. and Darby, S.C. (1987). Modelling the relative and absolute risks of radiation-induced cancers. *Journal of the Royal Statistical Society A*, **150**, 83–118.

National Rivers Authority (1994). *Water Quality Objectives: Procedures used by the National Rivers Authority for the purpose of the Surface Waters (River Ecosystem) (Classification) Regulations 1994*. NRA, Bristol, March.

National Rivers Authority (1996). *The Worcestershire Stow Catchment, Severn Trent Region: Proposals for Statutory Water Quality Objectives*. NRA, Bristol, March.

Neil, B.C.J. (1993). The derived emission limit for tritiated hydrogen gas from the Darlington Tritium Removal Facility. *Journal of Fusion Energy*, **12**, 171–175.

Nyholm, N.E.I. (1995). Monitoring of terrestrial environmental metal pollution by means of free-living insectivorous birds. *Annali di Chimica*, **85**, 343–351.

Office of Air Quality Planning and Standards (1996). *EPA CFR Part 50: National Ambient Air Quality Standards for Ozone and Particulate Matter*. US EPA, Research Triangle Park, NC.

O'Hagan, A. (1995). *Elicitation of Hydrological Data*. HMIP Report TR-Z2–17.

O'Hagan, A. (1996). Eliciting expert beliefs in substantial practical applications. Nottingham Statistical Research Report 95–1.

O'Hagan, A. and Haylock, R.J. (1996). Bayesian uncertainty analysis and radiological protection. In Barnett, V. and Turkman, K.F. (eds) *Statistics for the Environment 3*. Wiley, Chichester.

O'Hagan, A. and Wells, F.S. (1993) Use of prior information to estimate costs in a sewerage operation. In Gatsonis, C., Hodges, J.S., Kass, R.E. and Singpurwalla, N.D. (eds) *Case Studies in Bayesian Statistics*, Springer-Verlag, New York, pp. 118–163.

Organization for Economic Cooperation and Development (1992). *Report of the OECD Workshop on the Extrapolation of Laboratory Aquatic Toxicity Data to the Real Environment*, OECD Environment Monographs No. 59. OECD, Paris.

Osenberg, C.W., Schmitt, R.J., Holbrook, S.J., Abusaba, K.E. and Flegal, A.R. (1994). Detection of environmental impacts – natural variability, effect size, and power analysis. *Ecological Applications*, **4**, 16–30.

Paladino, O. (1994). Transition to chaos in continuous processes: Applications to waste water treatment reactors. *Environmetrics*, **5**, 57–70.

Pan, Z.Q. (1995). Radiation exposures caused by the nuclear industry in China. *Radiation Protection Dosimetry*, **62**, 245–254.

Peterson, C.H. (1993). Improvement of environmental impact analysis by application of principles derived from manipulative ecology – lessons from coastal marine case histories. *Australian Journal of Ecology*, **18**, 21–52.

Piegorsch, W.W. (1994). Empirical Bayes calculations of concordance between end points in environmental toxicity experiments. *Environmental and Ecological Statistics*, **1**, 153–162.

Piegorsch, W.W., Smith, E.P., Edwards, D. and Smith, R.L. (1996). Statistical advances in environmental science. Submitted for publication.

Pierce, D.A., Schimizu, Y., Preston, D.L. *et al.* (1996). Studies of the mortality of A-bomb survivors, Report 12, Part 1. Cancer 1950–1990. *Radiation Research*, **146**, 1–27.

Redmond, G. K. and Mazumdar, S. (1993). Cold oven washers. *International Statistical Review*, **61**, 207–221.

Reddy, P.J., Barbarick, D.E. and Osterburg, R.D. (1995). Development of a statistical model for forecasting episodes of visibility degradation in the Denver metropolitan area. *Journal of Applied Meteorology*, **34**, 616–625.

Risk Assessment Forum (1995). *The Use of the Benchmark Dose Approach in Health Risk Assessment*. Environmental Protection Agency, Washington, DC.

Roberts, E.A. (1993). Seasonal cycles, environmental change and BACI designs. *Environmetrics*, **4**, 209–231.

Rosenbaum, B.J., Strickland, T.C. and McDowell, M.K. (1994). Mapping critical levels of ozone, sulfur dioxide and nitrogen dioxide for crops, forests and natural vegetation in the United States. *Water, Air and Soil Pollution*, **74**, 307–319.

Scharer, B. (1995). Recent developments in technologies and policies in Germany to control acid deposition. *Water, Air and Soil Pollution*, **85**, 1885–1890.

Schonhofer, F and Pock, K (1995). Incorporation of tritium from wrist watches. *Journal of Radioanalytical and Nuclear Chemistry*, **197**, 195–202.

Schroeter, S.C., Dixon, J.D., Kastendiek, J. and Smith, R.O. (1993). Detecting the ecological effects of environmental impacts – a case-study of Kelp Forest invertebrates. *Ecological Applications*, **3**, 331–350.

Schwartz, J. (1994). Nonparemetric smoothing in the analysis of air-pollution and respiratory illness. *Canadian Journal of Statistics*, **22**, 471–487.

Shackley, S., Young, P., Parkinson, S. and Wynne, B. (1996). Uncertainty, complexity and concepts of good science in climate change modelling: are GCMs the best tools? Submitted to *Climate Change*.

Seber, G.A.F., and Thompson, S.K. (1994). Environmental adaptive sampling. In Patil, G.P. and Rao, C.R. (eds) *Handbook of Statistics Volume 12: Environmental Statistics*. North-Holland/Elsevier, New York, pp. 210–220.

Slob, W. (1993). Modeling long-term exposure of the whole population to chemicals in food. *Risk Analysis*, **13**, 525–529.

Slob, W. (1996). Personal Correspondence.

Slob, W. and Krajnc, E.I. (1994). Interindividual variability in modeling exposure and toxicokinetics: a case study on cadmium. *Environmental Health Perspectives*, **102**, 78–81.

Smith, A.F.M. (1993). *A review of probabilistic and statistical issues in quantitative risk analysis for radioactive waste repositories. Part 1: Overview.* Report DOE.HMIP/PR.93.07. DOE, London.

Smith, R.I. (1996). Personal Correspondence.

Smith, R.L. and Huang, L.-S. (1994). *Modeling High Threshold Exceedances of Urban Ozone, Technical Report 6.* National Institute of Statistical Sciences, Research Triangle Park, NC.

Solov, A.R., Gaines, A.G. (1995). An empirical Bayes approach to monitoring water quality. *Environmetrics*, **6**, 1–5.

Strickland, T.C., Holdren, G.R., Ringold, P.L., Bernard, D., Smythe, K. and Fallon, W. (1993). A national critical loads framework for atmospheric deposition effects assessment. 1. Method summary. *Environmental Management*, **17**, 329–334.

Thompson, B.G.J. and Sagar, B. (1993). The development and application of integrated procedures for post-closure assessment, based upon Monte Carlo simulation: the probabilistic system assessment approach. *Reliability Engineering and System Safety*, **42**, 125–160.

Thompson, B.G.J., Smith, R.E. and Porter, I. (1996). Some issues affecting the regulatory assessment of long-term post-closure risks from underground disposal of radioactive wastes. Paper presented at the PSAM III Conference, Crete, June.

Tirabassi, T. and Rizza, U. (1994). An analytical air pollution model for complete terrain. *Environmetrics*, **5**, 159–165.

Topfer, K. (1994). The challenge of integrating the environment into policy-making. *Marine Pollution Bulletin*, **29**, 266–269.

Underwood, A.J. (1993). The mechanics of spatially replicated sampling programs to detect environmental impacts in a variable world. *Australian Journal of Ecology*, **18**, 99–116.

von Holstein, C.-A.S.S. and Matheson, J.F. (1978). *A Manual for Encoding Probability Distributions*, SRI Report AD-A 092259. SRI International, California.

Wadsworth, R.A. and Brown, M.J. (1995). A spatial decision-support system to allow the investigation of the major impact of emissions from major point sources under different operating policies. *Water, Air and Soil Pollution*, **85**, 2649–2654.

Walker, S.R. (1992). *Procedures for the Elicitation of Expert Judgements in the Probalisitc Risk Analysis of Radioactive Waste Repositories: An Overview.* DOE Report DOE/HMIP/RR.92/118. DOE, London.

Warn, T. (1996). Setting environmental standards. Personal Correspondence.

Warwick, J.J. and Roberts, L.A. (1992). Computing the risks associated with wasteload allocation modeling. *Water Resources Bulletin*, **28**, 903–915.

Wasserman, H.J. and Klopper, J.F. (1993). Analysis of radiation doses received by the public from I-131 treatment of thyrotoxic outpatients. *Nuclear Medicine Communications*, **14**, 756–760.

Weisel, C.P., Cody, R.P. and Lioy, P.J. (1995). Relationship between summertime ambient ozone levels and emergency department visits for asthma in central New Jersey. *Environmental Health Perspectives*, **103**, 97–102.

Wijedasa, H.A. and Kemblowski, M.W. (1993). Bayesian decision analysis for plume interception wells. *Ground Water*, **31**, 948–952.

Wyzga, R.E. and Folinsbee, L.J. (1995). Health effects of acid aerosols. *Water, Air and Soil Pollution*, **85**, 177–188.

Young, P., Parkinson, S. and Lees, M. (1996). Simplicity out of complexity in environmental modelling: Occam's razor revisited. *Journal of Applied Statistics*, **23**, 165–210.

Zach, R., Amiro, B.D., Davis, P.A., Sheppard, S.C. and Szekeley, J.G. (1994). Biosphere model for assessing doses from nuclear waste disposal. *Science of the Total Environment*, **156**, 217–234.

Zeevaert, T., Volckaert, G. and Vancecasteele, C. (1995). A sensitivity study of the SCK–Center–DOT–CEN biosphere model for performance assessment of near-surface repositories. *Health Physics*, **69**, 243–256.

Index

Acid rain 6, 59
Aerosol 65
Air pollution 2, 3, 52, 54, 55, 56, 59,
    79–82, 86, 87
Air Quality 69
ALARA, see As low as reasonably
    achievable
ALARP, see As low as reasonably
    practical
Aluminium 54
Alzheimer's 54
Ammonia 2, 23, 28, 77
Arsenic 59
Asbestos 64
As low as reasonably achievable
    (ALARA) 37, 62, 83, 92, 93
As low as reasonably practical (ALARP)
    82, 83, 85
Asthma 24, 56
Atomic bombs 45, 48, 60
Australia, see New South Wales
    Environmental Protection Agency,
    see Environment Council
Average, see Sample average

BACI designs 53–54, 57
Bathing water 2, 3, 7, 24, 69, 86, 87
    quality 3, 7, 86, 87
Bayesian analysis 47, 54, 55
    empirical Bayes methods 55
Benchmark
    dose (BMD) 72
    level 55

Benefit 17, 18, 20, 35, 89, 90, 96
Benefit of the doubt 31, 78
Benzene 65
Best available technology not entailing
    excessive cost (BATNEEC) 83, 37,
    38, 62, 92, 93
Best practical means (BPM) 82, 83
Binomial distribution 81
Birds 59, 79
BPEO 93
BS 5750 15

Cadmium 59
Cancer 45–46
Carbon cycle 57
Carcinogens 71
Catalytic converter 22
Causal links 44
Chain from cost to benefit, see Cost
    benefit chain
Chaos 56
Chernobyl 64
Classification of pollution 5, 50–51
Climate change 57
Coastal sea water 3, 57
Combining information 51, 60–61
Complexity 22, 90
Compliance 2, 22, 80,
    criterion 76, 91
    monitoring 7, 10, 87
    procedure 77, 78, 80
    testing 7–8, 10, 80, 84
Concentration of pollutant 3

Confidence
   intervals 27, 28, 30, 47
   limit 72
Consistency of standards 35
Contact location 19, 22, 33, 41–42, 90
Contaminated land 66, 67
Copper 76
Cost benefit
   analysis 36, 94, 96
   balanced 7, 15, 17, 18, 20, 41, 60, 77,
      78, 79, 83, 87
   chain 7, 41, 82, 84
   studies 7
Costs 7, 14–15, 17, 18, 20, 35, 89, 90,
      96
   of sampling 53
Critical group 4, 57–58, 84, 61, 92, 93
Critical level 51, 56, 58–60, 61, 69, 70,
      92, 93
Critical load 51, 56, 58–60, 61, 69, 70,
      92, 93
Crops 58, 59
Culture change 97

Decision theory 13, 35, 78, 79, 94
Department of the Environment, UK 6,
      11, 59, 68, 82, 86
Deterministic model 43, 56–57
Diesel fuel 14
Dose-response 71
   curve 72
   relationship 45–46, 51, 54–55
Drinking water 56

E Coli 2
Edinburgh International Workshop 18,
      30
Effect of pollutant 1–2, 4, 6, 8–9,
      41–61
   link with level 7–9, 41–61
EMAP, US 52
Empirical model 43
Entry location 19, 22, 33, 41–42, 90
Environment Agency (EA) 31, 69, 77, 79,
      83, 84
Environment Council (Australia and New
      Zealand) 63, 68
Environmental impact analysis (EIA) 51,
      57
Environmental mapping 59

Environmental pollution 1
   environmental pollution standards, see
      Standards for environmental
      pollution
Environmental Protection Agency US 37,
      38, 55, 63, 64, 65, 66, 70, 79–82
Environmental statistics 10, 51, 97
Estimates 45, 47, 61
European Union (EU) 7
Expert opinion 45, 48, 52, 61
   elicitation 47, 48, 84
Expert Panel on Air Quality Standards
      65
Extrapolation 48, 51
Extrapolation factors 60–61
Extreme values 14, 43, 51, 53, 56, 60, 86,
      92

Factory emissions 6
Faecal coliforms 6, 68, 86
Fertilizers 19
Fish 56, 57, 58, 78, 79
Flexibility 22, 90
Forests 59
   decline 69

Gaseous pollutants 52, 59
Generalised linear model (GLM) 55, 56
Germany
   air quality 65
Goldminers 58
Greenhouse gases 2
Groundwater 64

Hazardous waste 64
Her Majesty's Inspectorate of Pollution
      (HMIP) 82, 83
Hindsight 89
Hospitals 56
Hybrid model 44–49
Hydrology 56

Ideal standard 17, 48, 76, 77, 78, 79, 86,
      87, 90
   formulation 23, 24
   need to assess compliance 91
   positions 21
Impact of pollutant 4, 57
   impact measure 4
Industrial effluent 78

Iodine 2
ISO 9000 14

Kuwait 65

Lead 59
Leukaemia 53
Level of pollutant 1, 48, 76, 80, 86
    link with effect 7, 41–46, 86
    setting level in standard 77, 80,
        81–82
Levels of standard 35, 90, 91, 94, 96
    guiding principles 37
    role of statistical verification standard
        94
Linear model 55
Location of pollutant 3
    contact location 19, 22, 41–42
    entry location 19, 22, 41–42
Logistic regression 65
Logit 55
Lognormal distribution 78
London Airport 36
Lowest-observed-effect concentration
    (LOEC) 55, 57

Manure 86
Mean, see Sample mean
Measurement errors 2, 8, 9
Mechanistic model 43–44, 56–57, 60, 67,
        69, 73, 92, 95
    compartmental approach 57, 58
Medium 2, 86
Mercury 58, 59
Meta-analysis 60
Meteorology 56
Model
    deterministic, see Deterministic model
    empirical, see Empirical model
    hybrid, see Hybrid model
    mechanistic, see Mechanistic model
    statistical, see Statistical model
    validation 48–49
Monte Carlo approach 49–50, 64, 78
Multiple standards 33, 91
    consistency 34
Multivariate data 4, 42
Multivariate statistical methods 56
    canonical variates 56
    cluster analysis 56

pattern recognition 56
principal components 56

National Radiological Protection Board
        45
National Rivers Authority (NRA) 77, 78
Natural variability 2, 8, 41, 42
Netherland's Health Council 73
Neural nets 56
New South Wales Environmental
        Protection Agency 63, 66
New Zealand, see Environment Council
Nitrates 2, 19, 20, 24, 33, 56, 86
Nitrogen dioxide 3, 87
Non-parametric statistical methods 55, 56
    smoothing 56
    Wilcoxon tests 56
No observed effect-level (NOEL) 72
Normal distribution 81
Nuclear waste 6, 11, 77, 82–86
    disposal 5, 11, 77, 82–86
    power 9

Objective: standard as 18
Observation errors 2, 8
OECD 60
Oxygen 77, 87
Ozone 3, 54, 56, 59, 64, 65, 71, 74, 77,
        79
    US standard 79–82

Paper pulp 87
Parameter 44, 47
Particulate matter 9, 24, 59, 65
Percentile 24, 25, 28, 35
Pesticide 73
Pharmacokinetics 57
Plants 59
Poisson distribution 81
Pollutant 2
    classification scheme, see Classification
        of pollution
    concentration, see Concentration of
        pollutant
    effect, see Effect of pollutant
    impact, see Impact of pollutant
    level, see Level of pollutant
    taxonomy, see Classification of
        pollution
    threshold, see Threshold of pollutant

Pollutant-effect
    relationship 7–9, 25, 33, 40, 41–61,
        74, 76, 87, 90, 92, 94
    statistical analysis 50–56
    surveys 50–56
Position of standard 96
    contact location, see Contact location
    definition 19
    entry location, see Entry location
    flexibility and complexity, see
        Complexity, and Flexibility
    ideal and realisable standards 21
Posterior probability interval 27
Precautionary principle 18, 37, 39, 62, 69,
        92, 93
Probalilistic risk analysis 95
Probability 96
    assessment 2
    degree of belief interpretation 13
    density function 12
    distribution 11–12, 34, 46–47, 49
    distribution function 11–12
    frequency interpretation 12–13
    fundamental role 10
    interpretation 12–13
Probit 55
Process control 15
Prudent reduction 37, 62, 93

Quality assurance 81

Radiation exposure 58, 82–86
Radioactive contamination 6, 9, 45–46,
        54, 57
Radiological protection 45–46
Range, see Sample range
Realizable standard 17, 21, 86, 87, 90
    on actions 30
    over-prescriptive 30
    relation to pollutant level 91
Regression 43, 45, 55
Reservoirs 56
Respiratory illness 9, 23, 24, 54, 56
    asthma 24, 56, 65
Return period 56
Risk 11, 13, 84, 92
    analysis 13, 67, 84, 92
    assessment 49–50, 84
    Monte Carlo method 49–50, 84
    perception 38

probabilistic 49–50, 85
River waste 3, 4
River water 3, 4, 20, 52, 59, 76, 77
    quality 3, 6, 52, 55
Royal Commission on Environmental
        Pollution 63, 68
Royal Statistical Society 95

Safe level 37, 38, 92
Safety factor 71
Sample 3
    average 3, 59, 82
    data 42
    mean 3, 82
    median 43, 86
    quality of 33
    range 3
Sample statistic 3, 7–8, 10
Sample unit 3
Sampling costs 8
Sampling methods 47, 48, 51, 52–53, 57
    adaptive 52–53
    composite 52–53
    multi-phase 53
    networks 53
    random 52
    ranked-set 47, 52–53
    'snowballing' 52
    uncertainty due to 26
Sea-floor sediment 1
Sea water 3
    quality 3
Sensitivity analysis 50, 84
Sewage 6, 14, 53, 58
Soil pollution 58, 59
Soil remediation 53
Spatial effects 2, 9, 42, 57, 79
    spatial autocorrelation 54
    spatial clustering 54, 55
SPRUCE 5, 50, 51
Standard error 47
Standards for environmental pollution
    as 'mandatory requirements' 1
    as 'requirements' 1
    as 'set of preferred actions' 1
    expressed in terms of level, see Level
        of pollutant
    ideal standard, see Ideal standard
    kinds of standards 6–7
    overview 5

position set at 42
realizable standard, *see* Realizable
    standard
set on effects 82
setting 5–8, 15–16, 83–84
statistical verification 48, 76, 77, 78,
    79, 83, 85, 86
Statistical
    methods 2, 42
    model 92
Statistical analysis 13, 47–49
    estimates 44–45, 52–53
    inference 13, 26
    methods 2, 7, 10, 14, 16, 42–43
    model 10, 14, 42, 42–43, 60, 92
    tests 44, 52
Statistically verifiable ideal standard 18, 27,
        28, 63, 74, 76, 78, 79, 80, 86, 97
    consistency of standards, *see*
        Consistency of standards
    contrast with ideal and realizable
        standards 29, 86
    example 28
    implementation issues 32–33
    statistical quality 30
Statistician, role of 14–15, 47, 49–50, 86
Statutory Water Quality Objectives
    (SWQOs) 77, 78
Storms 56
Subject group 4, 8, 41–42
Sulphur products 6, 59, 87
    damage to buildings 59
Surrogate measurements 48
Survival analysis 54

Temporal effects 2, 9, 42, 52, 54, 79, 80
    see Time series
Threshold of pollutant 3, 7, 56, 59
Time series 54
    autoregressive models 54
    Box-Jenkins forecasting 54
    Kalman filter methods 54
    non-linear 54
Tourism 78
Toxic effects 47–48, 55, 57
Toxicity 60, 71
Toxicokinetic model 73
Treaty of Maastricht 39

Uncertainty about parameters 44, 47

Uncertainty analysis 50, 84, 95
Uncertainty and variation 2
    in costs and benefits 33–34
    in current standards 76–87
    in levels of exposure 8
    in models 43–47, 95
    in pollutant-effect relationships 8–9,
        41–61
    in setting standards 15–16, 24, 63,
        76–87, 89, 93, 95–96
    intrinsic 42, 46–47, 63
    measurement errors 2, 8, 9
    natural variability 2, 8
    observation errors 2, 8
    random 8, 9
    random errors 42, 44–45
    recognition of 23–24, 38–39, 97
    representation 10–15
    sample variation 9–10, 21, 26, 63
    scientific 8–9
    spatial effects 2, 9, 42, 57, 63, 79
    temporal effects 2, 9, 42, 52, 54, 79,
        80
United States
    see Environmental Protection Agency
        US
    Food and Drug Administration 73
    Office of Air Quality and Planning
        Standards 65
Utility 36, 92

Variability, *see* Uncertainty and variation
Variation, *see* Uncertainty and variation

Warn-Brew method 78
Waste disposal 58
Waste water 78
    EU directive 28, 29, 30, 78
Water
    bathing, *see* Bathing water
    coastal, *see* Coastal sea water
    distribution 14–15
    drinking, *see* Drinking water
    pollution 42, 86, 87
    river, *see* River water
    sea, *see* Sea water
Weak links 47–48
World Health Organisation 64, 67

Zinc 77